MONTE CARLO METHODS FOR ELECTROMAGNETICS

T0199370

MONTE CARLO METHODS FOR ELECTROMAGNETICS

Matthew N. O. Sadiku

CRC Press
Taylor & Francis Group
Boca Raton London New York

CRC Press is an imprint of the
Taylor & Francis Group, an **informa** business

MATLAB® is a trademark of The MathWorks, Inc. and is used with permission. The MathWorks does not warrant the accuracy of the text or exercises in this book. This book's use or discussion of MATLAB® software or related products does not constitute endorsement or sponsorship by The MathWorks of a particular pedagogical approach or particular use of the MATLAB® software.

CRC Press
Taylor & Francis Group
6000 Broken Sound Parkway NW, Suite 300
Boca Raton, FL 33487-2742

First issued in paperback 2019

ISBN-13: 978-1-4398-0071-3 (hbk)
ISBN-13: 978-1-4398-0071-3 (hbk)

Dedication

Dedicated to my brothers

Moses, David, James, and Julius

Contents

Preface

The development of Monte Carlo methods should be regarded as an integral part of the trend in recent years to supplement experimental and analytic methods with numerical techniques. Numerical techniques have become the standard tools for solving practical, complex electromagnetic problems. There are five popular numerical techniques that are commonly used in the electromagnetic community: the method of moments, the finite difference methods, the finite element methods, the transmission-line-modeling matrix, and the Monte Carlo methods. In the past, new beginners in the field of numerical computation in electromagnetics had to painstakingly dig through the literature to discover these techniques. To alleviate this problem, monographs have been written on each of the methods with the exception of the Monte Carlo methods (MCMs). This is perhaps due to the fact that there is no other person in the field besides the present author who has done a lot of work on the MCMs. To meet the need, I have decided to put my work, along with that of others, into a monograph.

This book is an attempt to present Monte Carlo techniques in a systematic way. It is written for researchers and practitioners in the field of simulation in general and electromagnetic computation in particular. The book is ideal for electronics and electrical engineers, students, and researchers who want a clear understanding of the MCMs and their applications to electromagnetic computation. The background requirements needed to fully understand the text are satisfied by the knowledge acquired in an introductory course on numerical analysis that includes finite difference method.

Although the book is a monograph, examples and end-of-chapter problems are included so that it could be used as a graduate introductory textbook on the applications of Monte Carlo techniques in electromagnetics. The writing style of the author makes the concepts so clear and simple that even undergraduate students can benefit from the book. The book can also be used for a short course on Monte Carlo techniques.

The book is divided into 11 chapters. The first chapter is on introductory matters such as the historical background and a review of electromagnetic theory. Chapter 2 addresses probability and statistics. Chapter 3 provides an introduction to finite difference method. Chapter 4 deals with fixed random walk MCM, while Chapter 5 is on floating random walk. Chapter 6 covers the Exodus method and applies the method to Laplace's and Poisson's equations. Chapter 7 contains Monte Carlo techniques for handling Neumann problems, while Chapter 8 deals with whole field computation using the Markov chain. Chapter 9 applies MCM to time-varying diffusion problems, while Chapter 10 is concerned with wave scattering due to random rough surfaces. Chapter 11 covers multidimensional integration.

With the exception of a few subroutines in Fortran, all computer programs are in MATLAB®.

MATLAB is a trademark of the MathWorks, Inc. and is used with permission. The MathWorks does not warrant the accuracy of the text or exercises in this book. This book's use or discussion of MATLAB software or related products does not constitute endorsement or sponsorship by The MathWorks of a particular pedagogical approach or particular use of the MATLAB software. For product information, please contact:

The MathWorks, Inc.
3 Apple Hill Drive
Natick, MA 01760-2098 USA
Tel: 508 647 7000
Fax: 508-647-7001
E-mail: info@mathworks.com
Web: www.mathworks.com

I am delighted to acknowledge the contributions made to my understanding of this subject by a number of colleagues. Specifically, I am indebted to Dr. Keming Gu, who developed the idea in Chapter 6, and to Dr. Sudarshan Nelatury of Penn State University for writing part of Chapter 3. I owe special thanks to Dr. Kendall Harris, our college dean at Prairie View A&M University, and Dr. John O. Attia, our head of department for their support. I thank my daughters, Ann and Joyce, for helping with the figures. I also wish to thank my wife, Kikelomo, for her support and prayers.

Matthew N. O. Sadiku
Prairie View, Texas
June 2, 2008

Author Biography

Matthew N. O. Sadiku is presently a professor at Prairie View A&M University. He is the author of over 180 professional papers and over 30 books including *Elements of Electromagnetics* (Oxford, fourth edition, 2007), *Fundamentals of Electric Circuits* (McGraw-Hill, third edition, 2007, with C. Alexander), *Numerical Techniques in Electromagnetics with MATLAB®* (CRC, third edition, 2009), and *Metropolitan Area Networks* (CRC Press, 1995). Some of his books have been translated into Korean, Chinese (and Chinese Long Form in Taiwan), Italian, Portuguese, and Spanish. He was the recipient of the 2000 McGraw-Hill/Jacob Millman Award for outstanding contributions in the field of electrical engineering. His current research interests are in the areas of numerical modeling of electromagnetic systems and computer communication networks. He is a registered professional engineer and a senior member of the Institute of Electrical and Electronics Engineers (IEEE).

1

Introduction

Computations: No one believes them, except the person who made them.
Experiments: Everyone believes them, except the person who made them.

—**B. Munk**

Monte Carlo Methods (MCMs) are a modern tool for the numerical analysis of electromagnetic problems. This chapter provides a justification for, and a brief historical background of, MCMs. It also presents a brief review of electromagnetic theory as a background for the remaining chapters.

1.1 Why Monte Carlo?

Until the 1940s, physical problems were classified into two categories: theoretical and experimental. With the advent of the modern digital computer, a third category has emerged: computer simulation. Computer simulation should not be regarded as replacing the first two traditional approaches to solving physical problems; rather, it complements both theoretical and experimental analyses.

Besides the fact that a high degree of ingenuity, skill, and effort are required for theoretical or analytic solutions, only a narrow range of practical problems can be investigated, owing to the complex geometries defining the problems. Also, theoretical analysis often involves some mathematical approximations that are not justifiable from first principles. On the other hand, experiments are expensive, time consuming, sometimes hazardous, and do not often allow much flexibility in parameter variation. Computer simulation offers the freedom and flexibility to adjust various parameters of the system. It has the advantage of allowing the actual work to be carried out by operators without a knowledge of higher mathematics or physics, with a resulting economy of labor on the part of the highly trained personnel. Of course, a number of factors dictate which of the three approaches (theoretical, experimental, or computer simulation) should be used in solving a given physical problem.

Computer simulation of physical problems has led to the development of various numerical techniques. These techniques have become the standard modern tools for solving practical, complex electromagnetic problems. Popular numerical techniques that are commonly used in the electromagnetic community include the method of moments, finite difference methods, finite element methods, the transmission-line-modeling matrix, and Monte Carlo methods (MCMs) [1]. Whereas the other numerical techniques are deterministic, MCMs are nondeterministic, probabilistic or stochastic. What is nondeterministic about MCMs is the approach, which involves random numbers, not the system or problem involved. The system modeled in Monte Carlo simulation may or may not be deterministic.

It is not true that Monte Carlo methods (such as fixed and floating random walks) generally do not provide accurate solutions. As we shall see later in this text, some MCMs (such as the Exodus method and the Markov chain MCM) do provide accurate solutions and are not subject to randomness as are other MCMs. Although Monte Carlo simulation is a slow and costly technique, it does have some advantages and is often used when all else fails. The major advantages of MCMs include the following [2,3]:

1. The methods are easy to understand and apply. They do not rely heavily on mathematical abstractions that require the assistance of an expert. Hence, they can be used by many more individuals.

2. Algorithms are simple to develop, allowing one to solve problems with complicated solution regions.

3. Computer code does not require large memory even for multidimensional problems.

4. It is possible to obtain the solution directly at an arbitrary point without the whole field computation.

5. It enables one to solve problems with stochastic parameters.

6. Statistical error can be provided simultaneously with the solution.

1.2   Historical Background

Computer simulation emerged as a numerical problem-solving approach during World War II when the American mathematicians John von Neumann and Stanislaw Ulam successfully used MCMs in their work on neutron diffusion problems associated with the development of the atomic bomb at the Los Alamos scientific laboratory. Although proposals for using MCMs had been made even earlier by notable scientists such as Einstein, Lord Rayleigh, Courant, and many others, only a few serious calculations could be performed because of the tedious, time-consuming process involved in generating random variables by hand. However, with computers, MCMs have been used with great success in solving diverse scientific and engineering problems.

The term *Monte Carlo* evolved from the association of random numbers with gambling at the casinos in the principality of Monte Carlo in Monaco. It was introduced in 1944 by von Neumann and Ulam as a secret code name for their work. The publication of the work by Metropolis and Ulam [4] in 1949 sparked a rapid growth and systematic development of MCMs. The last four decades have seen an intense effort to develop better Monte Carlo algorithms.

The phrase "Monte Carlo method" has been broadly used. The method may be regarded as a means of solving problems numerically through sampling experiments. It is based on the numerical realization of a model of the physical phenomena under consideration. Starting from a description of the model, it uses random numbers to construct the appropriate probability, which is then weighted with the various generated states of the system. Thus, the Monte Carlo method may be defined as a numerical method of solving a physical problem by the simulation of random variables [5].

1.3 Applications of MCMs

MCMs have now found numerous applications in many branches of science, engineering, business, and other disciplines. Much of the application of MCMs in electromagnetics (EM) has been in the area of potential theory. The connection between potential theory and Brownian motion was first shown in 1944 by Kakutani [6]. The resulting so-called probabilistic potential theory has been applied to problems in many disciplines such as structural analysis [7], heat transfer [8], electrostatics, and optimization [9]. It should be noted that the MCMs described in this text are special cases of the more general Monte Carlo simulation.

MCMs also find fascinating widespread applications in stochastic differential equations, boundary value problems [3], atmosphere and ocean optics [10,11], polymer science [12], semiconductor devices [13], statistical physics [14–16], and quantum field theory [17,18]. These scientific applications are not the subject of this text, although some ideas from these areas are applicable to EM problems.

MCMs have some unique, exciting features that are worthy of note. First, they yield exact information (besides some statistical errors, which can be minimized, at least in principle) on system models. In general, system models are more precisely characterized in Monte Carlo simulation than in theoretical analysis. Thus, Monte Carlo simulations (or computer simulations in general) are often used to check the accuracy and range of validity of some approximations made in the analytic treatment of a model. Second, the Monte Carlo solution of a problem is closer in spirit to physical experiments than to classical numerical techniques. Third, Monte Carlo simulation is interdisciplinary in nature. For example, MCMs developed for the analysis of heat transfer in mechanical engineering have become useful for Monte Carlo solution of EM problems. Thus, progress made in one area may trigger better understanding of a rather different problem in a different discipline.

1.4 Review of Electromagnetic Theory

A detailed treatment of the physical laws governing electromagnetic phenomena can be found in standard introductory EM texts such as [19,20]. Only a brief review will be presented here.

1.4.1 Maxwell's Equations

The whole subject of EM unfolds as a logical deduction from seven postulated equations, namely, Maxwell's four field equations and three medium-dependent equations.

The time-varying EM fields are governed by physical laws expressed mathematically as follows:

$$\nabla \bullet \mathbf{D} = \rho_v \qquad (1.1)$$

$$\nabla \bullet \mathbf{B} = 0 \qquad (1.2)$$

$$\nabla \times \mathbf{E} = -\frac{\partial \mathbf{B}}{\partial t} \qquad (1.3)$$

$$\nabla \times \mathbf{H} = \mathbf{J} + \frac{\partial \mathbf{D}}{\partial t} \qquad (1.4)$$

where
  $\mathbf{E}$ = electric field intensity in volts per meter (V/m)
  $\mathbf{D}$ = electric flux density in coulombs per square meter (C/m^2)
  $\mathbf{H}$ = magnetic field intensity in amperes per meter (A/m)
  $\mathbf{B}$ = magnetic flux density in webers per square meter (Wb/m^2)
  $\mathbf{J}$ = electric current density in amperes per square meter (A/m^2)
  ρ_v = electric charge density in coulombs per cubic meter (C/m^3)

Equations 1.1 to 1.4 are referred to as Maxwell's equations in differential (point or local) form. They are first-order linear coupled differential equations relating the vector field quantities to each other at any given point. Their equivalent integral (or global) forms are as follows:

$$\oint_S \mathbf{D} \bullet dS = \oint_v \rho_v dv \tag{1.5}$$

$$\oint_S \mathbf{B} \bullet dS = 0 \tag{1.6}$$

$$\oint_L \mathbf{E} \bullet dl = -\int_S \frac{\partial \mathbf{B}}{\partial t} \bullet dS \tag{1.7}$$

$$\oint_L \mathbf{H} \bullet dl = \int_S \left(J + \frac{\partial \mathbf{D}}{\partial t} \right) \bullet dS \tag{1.8}$$

where v is an arbitrary volume bounded by a closed surface S in Equations 1.5 and 1.6 and S is an arbitrary surface bounded by a closed path L in Equations 1.7 and 1.8. Note that the terms on the right-hand side of Equations 1.1 to 1.8 represent the sources of the fields.

In addition to these four Maxwell's equations, there are three medium-dependent equations:

$$\mathbf{D} = \varepsilon \mathbf{E} = \varepsilon_o \varepsilon_r \mathbf{E} \tag{1.9}$$

$$\mathbf{B} = \mu \mathbf{H} = \mu_o \mu_r \mathbf{H} \tag{1.10}$$

$$\mathbf{J} = \sigma \mathbf{E} \tag{1.11}$$

where
$\varepsilon_o \simeq \frac{10^{-9}}{36\pi}$ = permittivity of free space in farads per meter (F/m)
ε_r = dielectric constant (or relative permittivity) of the medium
ε = permittivity of the medium in farads per meter (F/m)
$\mu_o = 4\pi \times 10^{-7}$ = permeability of free space in henries per meter (H/m)
μ_r = relative permeability of the medium
μ = permeability of the medium in henries per meter (H/m)
σ = conductivity of the medium in siemens per meter (S/m)

Equations 1.9 to 1.11 are called *constitutive relations* for the medium in which the fields exist. Equations 1.5 to 1.11 form the seven postulated equations based on which EM theory unfolds.

1.4.2 Scalar and Vector Potentials

To avoid the difficulty of solving directly the vector fields associated with Maxwell's equations, it is sometimes useful to employ the electric scalar potential V and magnetic vector potential $\mathbf{A}$. The vector potential is used

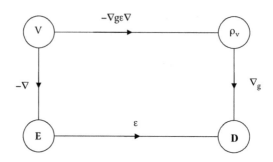

FIGURE 1.1
Relationship between V, **E**, and **D**.

when the solution region contains current densities, whereas the scalar potential is used in a domain devoid of currents. We now obtain these potentials from Maxwell's equations.

Under static conditions, the electric and magnetic fields can exist separately. For electrostatic fields, the appropriate Maxwell's equations in Equations 1.1 and 1.3 become

$$\nabla \bullet \mathbf{D} = \rho_v \tag{1.12}$$

$$\nabla \times \mathbf{E} = 0 \tag{1.13}$$

In terms of the electric potential V (in volts), **E** can be expressed as

$$\mathbf{E} = -\nabla V \tag{1.14}$$

The relationship between the potential V and vector fields **E** and **D** is portrayed in Figure 1.1. Combining Equations 1.12 and 1.14 gives Poisson's equation:

$$\nabla \bullet \varepsilon \nabla V = -\rho_z \tag{1.15}$$

or if ε is constant,

$$\nabla^2 V = -\frac{\rho_v}{\varepsilon} \tag{1.16}$$

When $\rho_v = 0$, Equation 1.15 becomes Laplace's equation:

$$\nabla \bullet \varepsilon \nabla V = 0 \tag{1.17}$$

or for constant ε

$$\nabla^2 V = 0 \tag{1.18}$$

Equations 1.15 and 1.17 are, respectively, Poisson's and Laplace's equations for inhomogeneous media, whereas Equations 1.16 and 1.18 are, respectively, Poisson's and Laplace's equations for homogeneous media.

For magnetostatic fields, Maxwell's equations (Equations 1.2 and 1.4) become

$$\nabla \bullet \mathbf{B} = 0 \tag{1.19}$$

$$\nabla \times \mathbf{H} = \mathbf{J} \tag{1.20}$$

In terms of the magnetic vector potential $\mathbf{A}$ (in Wb/m), we may express $\mathbf{B}$ as

$$\mathbf{B} = \nabla \times \mathbf{A} \tag{1.21}$$

Applying the vector identity

$$\nabla \times (\nabla \times \mathbf{F}) = \nabla(\nabla \bullet \mathbf{F}) - \nabla^2 \mathbf{F} \tag{1.22}$$

to Equations 1.19 to 1.21 and assuming that $\nabla \bullet \mathbf{A} = 0$, we get Poisson's equation for magnetostatic fields:

$$\nabla^2 \mathbf{A} = -\mu \mathbf{J} \tag{1.23}$$

When $\mathbf{J} = 0$, Equation 1.23 becomes Laplace's equation

$$\nabla^2 \mathbf{A} = 0 \tag{1.24}$$

Figure 1.2 presents the relationship between the vector potential $\mathbf{A}$ and vector fields $\mathbf{H}$ and $\mathbf{B}$.

1.4.3 Boundary Conditions

The material medium in which an EM field exists is usually characterized by its constitutive parameters σ, ε, and μ. The medium is said to be *linear* if σ, ε, and μ are independent of $\mathbf{E}$ and $\mathbf{H}$, or nonlinear otherwise. It is *homogeneous* if σ, ε, and μ are not functions of space variables, and inhomogeneous otherwise. It is *isotropic* if σ, ε, and μ are independent of direction (scalars), and anisotropic otherwise.

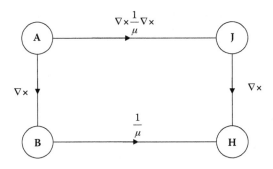

FIGURE 1.2
Relationship between $\mathbf{A}$, $\mathbf{B}$, and $\mathbf{H}$.

FIGURE 1.3
Interface between two media.

The boundary conditions at the interface separating two different media, 1 and 2, with parameters $(\sigma_1, \varepsilon_1, \mu_1)$ and $(\sigma_2, \varepsilon_2, \mu_2)$, as shown in Figure 1.3, are easily derived from the integral form of Maxwell's equations. They are

$$E_{1t} = E_{2t} \quad \text{or} \quad (\mathbf{E}_1 - \mathbf{E}_2) \times \mathbf{a}_{n12} = 0 \tag{1.25}$$

$$H_{1t} - H_{2t} = K \quad \text{or} \quad (\mathbf{H}_1 - \mathbf{H}_2) \times \mathbf{a}_{n12} = \mathbf{K} \tag{1.26}$$

$$D_{1t} - D_{2t} = \rho_s \quad \text{or} \quad (\mathbf{D}_1 - \mathbf{D}_2) \bullet \mathbf{a}_{n12} = \rho_s \tag{1.27}$$

$$B_{1t} - B_{2t} = 0 \quad \text{or} \quad (\mathbf{B}_1 - \mathbf{B}_2) \bullet \mathbf{a}_{n12} = 0 \tag{1.28}$$

where $\mathbf{a}_{n12}$ is a unit normal vector directed from medium 1 to medium 2; subscripts 1 and 2 denote fields in regions 1 and 2; and subscripts t and n denote tangent and normal components of the fields. Equations 1.25 and 1.28 state that the tangential components of $\mathbf{E}$ and the normal components of $\mathbf{B}$ are continuous across the boundary. Equation 1.26 states that the tangential component of $\mathbf{H}$ is discontinuous by the surface current density $\mathbf{K}$ on the boundary. Equation 1.27 states that the discontinuity in the normal component of $\mathbf{D}$ is the same as the surface charge density ρ_s on the boundary.

References

1. M. N. O. Sadiku, *Numerical Techniques in Electromagnetics*, 2nd ed., Boca Raton: CRC Press, 2001.
2. M. N. O. Sadiku and M. Ilyas, *Simulation of Local Area Networks*, Boca Raton: CRC Press, 1995, pp. 44, 45.
3. K. K. Sabelfeld, *Monte Carlo Methods in Boundary Value Problems*, Berlin/New York: Springer-Verlag, 1991, p. 2.
4. N. Metropolis and S. Ulam, The Monte Carlo method, *J. Amer. Stat. Assoc.* vol. 44, no. 247, 335–341, 1949.
5. I. M. Sobol, *A Primer for the Monte Carlo Method*. Boca Raton: CRC Press, 1994.
6. S. Kakutani, Two-dimensional Brownian motion harmonic functions, *Proc. Imp. Acad.* (Tokyo), vol. 20, 1944, pp. 706–714.
7. I. Elishakoff, *Probabilistic Methods in the Theory of Structures*, New York: John Wiley & Sons, 1983.

8. W. J. Minkowycz et al., *Handbook of Numerical Heat Transfer*, New York: John Wiley & Sons, 1988, pp. 673–722.
9. G. A. Mikhailov, *Optimization of Weighted Monte Carlo Methods*, Berlin: Springer-Verlag, 1992.
10. B. H. Sendov and I. Dimov (Eds.), *International Youth Workshop on Monte Carlo Method and Parallel Algorithms*, Teaneck, NJ: World Scientific Publishing, 1989.
11. G. I. Marchuk et al., *Monte Carlo Methods in Atmospheric Optics*, Berlin/New York: Springer-Verlag, 1980.
12. W. Bruns, I. Motoc, and K. F. O'Driscoll, *Monte Carlo Applications in Polymer Science*, Berlin/New York: Springer-Verlag, 1981.
13. M. Lundstrom, *Fundamentals of Carrier Transport*, Reading, MA: Addison-Wesley, 1990, pp. 201–233, 297, 298.
14. K. Binder, *Monte Carlo Method in Statistical Physics*, 2nd ed., Berlin/New York: Springer-Verlag, 1986.
15. K. Binder, *Applications of the Monte Carlo Method in Statistical Physics*, 2nd ed., Berlin: Springer-Verlag, 1987.
16. K. Binder and D. W. Heermann, *Monte Carlo Simulation in Statistical Physics: An Introduction*, Berlin: Springer-Verlag, 1992.
17. M. H. Kalos (Ed.), *Monte Carlo Methods in Quantum Problems*, Dordrecht, Holland: Reidel Publ. Co., 1984.
18. M. Suzuki (Ed.), *Quantum Monte Carlo Methods in Equilibrium and Nonequilibrium Systems*, Berlin: Springer-Verlag, 1987.
19. M. N. O. Sadiku, *Elements of Electromagnetics*, 4th ed., New York: Oxford University Press, 2007.
20. C. A. Balanis, *Advanced Engineering Electromagnetics*, New York: John Wiley & Sons, 1989.

2

Probability and Statistics

A pessimist is a man who looks both ways before crossing a one-way street.

—Laurence J. Peter

Applications of the Monte Carlo Methods (MCMs) to electromagnetic (EM) problems often require a mechanism for generating random *variates*—variables whose values obey a specified probability distribution. The generation of random variates requires random numbers. Thus, we begin this chapter by discussing how random numbers are generated by the computer and how the numbers can be tested for randomness before being applied to the Monte Carlo modeling of EM phenomena. The chapter also deals with the process of translating the random numbers into random variates and the statistical evaluation of error associated with the randomness in Monte Carlo simulations.

2.1 Generation of Random Numbers

Fundamental to Monte Carlo simulations is the need for sequences of numbers that appear to be drawn at random from a particular probability law. The method by which random numbers are generated is often called the *random number generator*. Various techniques for generating random numbers are discussed fully in [1–9]. A simple way of generating random numbers is by casting a dice with six faces numbered 1 to 6. Another simple way is to use the roulette wheel (similar to the "wheel of fortune"), shown in Figure 2.1. By turning the wheel and recording the number where arrow points, we obtain a sequence of random numbers such as

$$8, 0, 19, 11, 25, 5, 34, 16, 0, \ldots \tag{2.1}$$

These simple ways, however, will not generate enough numbers to make them truly random.

The almost universally used method of generating random numbers is to select a function $G(Z)$ that maps integers into random numbers. Select some

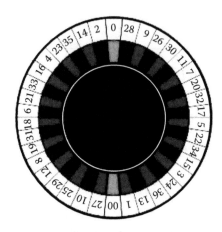

FIGURE 2.1
The roulette wheel.

guessed value Z_o, and generate the next random number as $Z_{n+1} = GZ_n$. The commonest function $G(Z)$ takes the form

$$G(Z) = (aZ + c) \bmod m \qquad (2.2)$$

where
Z_0 = a starting value or a seed ($Z_0 \geq 0$)
a = multiplier ($a \geq 0$)
c = increment ($c \geq 0$)
m = the modulus

The modulus m is usually 2^t for t-digit binary integers. For a 32-bit computer machine, for example, m may be 2^{32-1}. Here, Z_0, a, and c are integers in the same range as $m > a$, $m > c$, and $m > Z_0$. The desired sequence of random numbers Z_n is obtained from

$$Z_{n+1} = (aZ_n + c) \bmod m \qquad (2.3)$$

This is called a *linear congruential sequence*. For example, if $Z_0 = a = c = 7$ and $m = 10$, the sequence is

$$7, 6, 9, 0, 7, 6, 9, 0, \dots \qquad (2.4)$$

It is evident that congruential sequences always get into a loop; that is, there is ultimately a cycle of numbers that is repeated endlessly. The sequence in Equation 2.4 has a period of length 4. A useful sequence will, of course, have a relatively long period. The terms *multiplicative congruential method* and *mixed congruential method* are used by many authors to denote linear congruential methods with $c = 0$ and $c \neq 0$, respectively. Rules for selecting Z_0, a, c, and m can be found in [3,8]. The following three rules should be kept in mind:

1. The modulus m must be given a large value, for example, the word length of the computer. The increment c must be an odd integer and relatively prime to m; that is, c and m must have a common divisor. On a binary t-digit computer, it is recommended to select $m = 2^{t-1}$ for an efficient algorithm. For example, CYBER 205 uses $m = 2^{47}$.

2. The multiplier a must be prime relative to the modulus m. It is recommended that $a = 8r \pm 3$, where r is any positive integer, or we select a close to $2^{t/2}$, for example, $a = 2^{\text{integer }(t/2)} + 3$.

3. If $m = 2^t$ with $t \geq 3$ and $c = 0$, then the maximum cycle length is $p = 2^{t-2}$. This maximum is reached when $a = 4k + 1$ with odd integer k.

We have only discussed the linear congruential method of generating random numbers. Other methods (such as the midsquare generator, additive, and quadratic congruential generators) are discussed in Pooch and Wall [10].

In practice, we are usually interested in generating random numbers from the uniform distribution in the interval (0,1). These numbers will be designated by the letter U and are obtained from Equation 2.3 as

$$U = \frac{Z_{n+1}}{m} \tag{2.5}$$

Thus, U can only assume values from the set 0, $1/m$, $2/m$, ..., $(m-1)/m$. A set of uniformly distributed random numbers can be generated using the following procedure:

a. Select an odd number as a seed value Z_0.

b. Select the multiplier $a = 8r \pm 3$, where r is any positive integer and a is close to $2^{t/2}$. If $t = 31$, $a = 2^{15} + 3$ is a good choice.

c. Compute Z_{n+1} using either the multiplicative generator

$$Z_{n+1} = aZ_n \bmod m \tag{2.6}$$

or the mixed generator

$$Z_{n+1} = (aZ_n + c) \bmod m \tag{2.7}$$

d. Compute $U = Z_{n+1}/m$.

You only need the preceding procedure if your personal computer or computing facility does not have a random number generator. Most computer systems and languages (e.g., Fortran, Basic, C) are equipped with random number generators. For example, for a 32-bit computer, we may select $m = 2^{31} - 1 =$

```
SUBROUTINE RANDOM (ISEED,RN)

DOUBLE PRECISION ISEED, DEL, A

DATA DEL, A/2147483647.DO, 65539.DO/


IX = DMOD ( A*ISEED, DEL)

RN = IX/DEL

RETURN

END
```

FIGURE 2.2
Subroutine RANDOM.

$2,147,483,647$ and $a = 2^{16} + 3 = 65,539$. A subroutine to generate random numbers U according to Equations 2.5 and 2.6 is in Figure 2.2. By supplying a seed (e.g., ISEED = 1234), the subroutine provides one random number U per call in the main program. The seed is selected as any integer between 1 and m.

Random numbers based on mathematical relations in Equations 2.3 and 2.5 and produced by computer are not truly random. In fact, given the seed of the sequence, all numbers U of the sequence are completely predictable or deterministic. Some authors emphasize this point by calling such computer-generated sequences *pseudorandom numbers*. However, with a good choice of a, c, and m, the sequences of U appear to be sufficiently random in that they pass a series of statistical tests of randomness. They have the advantage over truly random numbers of being generated in a fast way and of being reproducible, when desired, especially for program debugging.

Example 2.1

    a. Using a linear congruential scheme, generate 10 pseudorandom numbers with $a = 573$, $c = 19$, $m = 10^3$, and seed value $Z_0 = 89$. Use these numbers to generate uniformly distributed random numbers $0 < U < 1$.

    b. Repeat the generation with $c = 0$.

Solution

    a. This is a multiplicative generator. Substituting $a = 573$, $c = 19$, $m = 1000$, and $Z_0 = 89$ in Equation 2.3 leads to

$$Z_1 = 573 \times 89 + 19 \ (\text{mod } 1000) = 16$$
$$Z_2 = 573 \times 16 + 19 \ (\text{mod } 1000) = 187$$
$$Z_3 = 573 \times 187 + 19 \ (\text{mod } 1000) = 170$$

$$Z_4 = 573 \times 170 + 19 \ (\text{mod } 1000) = 429$$

$$Z_5 = 573 \times 429 + 19 \ (\text{mod } 1000) = 836$$

$$Z_6 = 573 \times 836 + 19 \ (\text{mod } 1000) = 47$$

$$Z_7 = 573 \times 47 + 19 \ (\text{mod } 1{,}000) = 950$$

$$Z_8 = 573 \times 950 + 19 \ (\text{mod } 1000) = 369$$

$$Z_9 = 573 \times 369 + 19 \ (\text{mod } 1000) = 456$$

$$Z_{10} = 573 \times 456 + 19 \ (\text{mod } 1{,}000) = 307$$

Dividing each number by m = 1000 gives U as follows:

$$0.016, 0.187, 0.170, \ldots, 0.307$$

b. For $c = 0$, we have the mixed generator. Thus, we obtain

$$Z_1 = 573 \times 89 \ (\text{mod } 1000) = 997$$

$$Z_2 = 573 \times 997 \ (\text{mod } 1000) = 281$$

$$Z_3 = 573 \times 281 \ (\text{mod } 1000) = 13$$

$$Z_4 = 573 \times 13 \ (\text{mod } 1000) = 449$$

$$Z_5 = 573 \times 449 \ (\text{mod } 1000) = 277$$

$$Z_6 = 573 \times 277 \ (\text{mod } 1000) = 721$$

$$Z_7 = 573 \times 721 \ (\text{mod } 1000) = 133$$

$$Z_8 = 573 \times 133 \ (\text{ mod } 1000) = 209$$

$$Z_9 = 573 \times 209 \ (\text{mod } 1000) = 757$$

$$Z_{10} = 573 \times 757 \ (\text{mod } 1000) = 761$$

with the corresponding U as 0.997, 0.281, 0.013, ... , 0.761.

2.2 Statistical Tests of Pseudorandom Numbers

Pseudorandom numbers that are adequate for experimental purposes must satisfy some criteria. The quality of the random number generator is crucial to the success of the Monte Carlo simulations. A good generator must be capable of producing random numbers that have any of m discrete values which are equally likely to occur (i.e., the probability of a value's occurrence is $1/m$), and each new value is completely independent of any previous output of the generator. Statistically, this implies that the numbers are uniformly distributed, independent, and reproducible. It is also significant that the random numbers contain enough digits that generation of numbers on the interval (1,0) is sufficiently dense. A battery of statistical tests of uniform random number generators has been developed to reveal departures from independence and uniformity or to determine whether a series of numbers meets the criterion of randomness. For random numbers in the interval (0,1), a quick and simple test of the randomness is that the mean is 0.5. Other tests include moments, frequency, serial, gap, and poker tests [7,10,11]. Only the moments and frequency tests will be covered here.

2.2.1 Moments Test

This test provides an indication of the uniformity of the distributed pseudorandom numbers. Suppose $U_1, U_2, \cdots, U_N$ to be a sequence of uniformly distributed random numbers in the unit interval (0,1). The first, second, and third moments should be 1/2, 1/3, and 1/4, respectively, that is,

$$\frac{1}{N}\sum_{i=1}^{N} U_i = \frac{1}{2}$$

$$\frac{1}{N}\sum_{i=1}^{N} U_i^2 = \frac{1}{3} \qquad (2.8)$$

$$\frac{1}{N}\sum_{i=1}^{N} U_i^3 = \frac{1}{4}$$

where N is a very large number. In general, the k-th moment is

$$\frac{1}{N}\sum_{i=1}^{N} U_i^k = \frac{1}{k+1} \qquad (2.9)$$

The closer the calculated value of the moment to the expected value, the higher the quality of the random number sequence. Note that the simple test of randomness mentioned earlier corresponds with the first-moment test.

2.2.2 Frequency Test

Let $U_1, U_2, \cdots, U_N$ be a sample drawn from a population of uniformly distributed random numbers. The unit interval (0,1) is divided into n equal nonoverlapping divisions or subintervals of length $1/n$. The frequency test criterion uses the chi-squared statistic:

$$\chi^2 = \frac{n}{N}\sum_{i=1}^{n}\left(N_i - \frac{N}{n}\right)^2 \qquad (2.10)$$

where N_i is the number of random numbers in the i-th subinterval and $\sum_{i=1}^{n} N_i = N$.

To avoid distortion of results, we must select $n > 5$. The random numbers $U_1, U_2, \cdots, U_N$ are uniformly distributed and independent if

$$\chi^2 < \chi^2_{1-\alpha,n-1} \tag{2.11}$$

where α is the confidence level (say, 0.05 or 0.1), and $n - 1$ denotes the degree of freedom. In other words, we expect

$$\text{Prob}\left[\chi^2 < \chi^2_{1-\alpha,n-1}\right] = 1 - \alpha \tag{2.12}$$

The value of $\chi^2_{1-\alpha,n-1}$ can be obtained from the chi-square table in any standard statistical text or mathematics handbook, for example, [12–14]. When the criterion in Equation 2.11 is not met, the random number generation is said to be unsatisfactory and we reject the random numbers at the α-level of significance, which corresponds to a $100(1 - \alpha)\%$ confidence level.

A sample program for testing random number generators is provided in Bruns et al. [15].

Example 2.2

a. 5000 random numbers between 0 and 1 were produced by a random number generator. By dividing the interval into 10 equal subintervals (i.e., $0 \leq U < 0.1, 0.1 \leq U < 0.2, \ldots, 0.9 \leq U < 1.0$), the numbers located in each subinterval are counted as follows:

496 508 521 489 499 509 492 487 505 494

Use the frequency test to check whether the results are satisfactory or not.

b. Repeat Part a by dividing the interval into 5 equal subintervals (i.e., $0 \leq U < 0.2, 0.2 \leq U < 0.2, \ldots, 0.8 \leq U < 1.0$).

Solution

a. Using Equation 2.10,

$$\chi^2 = \frac{10}{5000} \sum_{i=1}^{n}\left(N_i - \frac{5000}{10}\right)^2$$

$$= 2 \times 10^{-3}[(496 - 500)^2 + (508 - 500)^2 + \cdots + (494 - 500)^2]$$

$$= 2 \times 10^{-3}[(-4)^2 + (8)^2 + (21)^2 + (-11)^2 + (-1)^2 + (9)^2 + (-8)^2 + (-13)^2 + (5)^2 + (-6)^2]$$

$$= 2.036$$

From the chi-square table, $\chi^2_{0.05,9} = 16.919$ for a confidence level of 0.05 with 9 degrees of freedom. Because $\chi^2 < \chi^2_{0.05,9}$, we accept the random numbers as satisfactory at the 0.05 level of confidence.

b. We first obtain the needed frequency by adding the numbers in two consecutive subintervals $0 \leq U < 0.1, 0.1 \leq U < 0.2, \dots , 0.9 \leq U < 1.0$) as

$$1004 \; 1010 \; 1008 \; 979 \; 999$$

$$\chi^2 = \frac{5}{5000} \sum_{i=1}^{n} (N_i - 1000)^2$$

$$= 10^{-3}[(4)^2 + (10)^2 + (8)^2 + (-21)^2 + (-1)^2]$$

$$= 0.622$$

For a confidence level of 0.05 with 4 degrees of freedom, $\chi^2_{0.05,4} = 11.070$, which is greater than $\chi^2 = 0.622$, showing that the random numbers are satisfactorily uniformly distributed.

2.3   Generation of Random Variates

Having discussed how to generate random numbers, we will now discuss various procedures for converting the random numbers into random variates.

It is often required in a Monte Carlo simulation to generate a random variable X from a given probability distribution $F(x)$. This can be accomplished using several techniques such as the *inverse transformation, rejection, rectangular approximation, composition, convolution,* and *look-up methods* [16–21]. The most commonly used techniques are the inverse transformation method and the rejection method.

2.3.1   Inverse Transformation Method

The inverse transformation method, otherwise known as the *direct method,* is the most straightforward technique for generating random variates from probability distribution functions. It basically entails inverting the cumulative probability function $F(x) = P[X \leq x]$ associated with the random variable X. The fact that $0 \leq F(x) \leq 1$ intuitively suggests that by generating random number U uniformly distributed over $(0,1)$, we can produce a random sample X from the distribution of $F(x)$ by inversion. Thus, to generate a random variable X with probability distribution $F(x)$, we set $U = F(x)$ and obtain

$$X = F^{-1}(U) \tag{2.13}$$

where X has the distribution function $F(x)$.

This technique for generating random variates assumes that the inverse transformation $F^{-1}(U)$, required in Equation 2.13, exists. This makes it suitable for generating random variates from exponential, gamma, uniform, and empirical distributions. However, there are several distributions for which $F^{-1}(U)$ cannot be found analytically, and the inverse transformation method cannot be used. A typical example is the normal distribution.

2.3.2 Rejection Method

This technique, also known as the *rejection–acceptance method*, can be applied to the probability distribution of any bounded variable. To apply the method, we let the probability density function of the random variable $f(x) = 0$ for $a > x > b$, and let $f(x)$ be bounded by M (i.e., $f(x) \leq M$) as shown in Figure 2.3. We generate random variates by taking the following steps:

1. Generate two random numbers (U_1, U_2) in the interval $(0,1)$.
2. Compute two random numbers with uniform distributions in (a, b) and $(0, M)$, respectively, that is,

   $$X = a + (b - a) U_1 \text{ (scale the variable on the } x\text{-axis)}$$

   $$Y = U_2 M \text{ (scale the variable on the } y\text{-axis)}$$

3. If $Y \leq f(X)$, accept X as the next random variate, otherwise reject X and return to Step 1.

Thus, in the rejection technique, all points falling above $f(x)$ are rejected, whereas those points falling on or below $f(x)$ are utilized to generate X

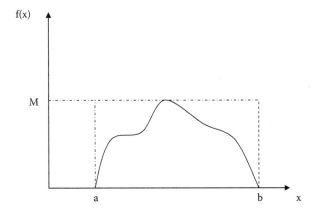

FIGURE 2.3
The rejection method of generating a random variate from $f(x)$.

through $X = a + (b - a)U_1$. The efficiency of the method is enhanced as the probability of acceptance in Step 3 increases.

2.4 Generation of Continuous Random Variates

Here, we discuss how the inverse transformation method is used in generating random variates for specific probability distribution functions. We will deal with three probability distributions that are important for our purpose.

It may be helpful at this point to define some important terms. A random variable is defined as a function that assigns a real number to each outcome in the sample space. The random variable X is characterized by its *probability density function* (pdf) $f(x)$ or its *cumulative probability function* (cdf) $F(x)$, also known as *probability distribution function*. Whenever it exists, the derivative of $F(x)$ is the probability density function $f(x)$, that is,

$$f(x) = \frac{dF}{dx} \tag{2.14}$$

or inversely

$$F(x) = \int_{-\infty}^{x} f(t)dt = P[X \le x] \tag{2.15}$$

The mean or expected value of a random variable X is defined as

$$\mu = E(X) = \int_{-\infty}^{\infty} xf(x)dx \tag{2.16}$$

The variance of a random variable is the weighted average of the values of $(x - \mu)^2$. In other words, it is a measure of the spread of the random variable's pdf about its mean. It is mathematically defined as follows:

$$\sigma^2 = V(X) = E[(X - \mu)^2] = \int_{-\infty}^{\infty} (x - \mu)^2 f(x)dx \tag{2.17}$$

or

$$\sigma^2 = V(X) = E[(X - \mu)^2]$$
$$= E(X^2 - 2\mu X + \mu^2) \tag{2.18}$$
$$= E(X^2) - \mu^2$$

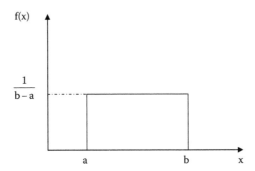

FIGURE 2.4
Uniform density function.

With these definitions, we can now discuss the three common probability distributions.

2.4.1 Uniform Distribution

This distribution, also known as *rectangular distribution,* is one in which the density is constant. It models random events in which every value between a minimum and maximum value is equally likely. Its graph is shown in Figure 2.4. Its characteristics are as follows:

$$f(x) = \frac{1}{b-a}, \quad a \le x \le b$$

$$F(x) = \int_a^x f(t)dt = \frac{x-a}{b-a}, \quad a \le x \le b \qquad (2.19)$$

$$E(X) = \frac{b+a}{2}$$

$$V(x) = \frac{(b-a)^2}{12}$$

A special uniform distribution for which $a = 0$ and $b = 1$ called the *standard uniform distribution,* is very useful in generating random samples from any probability distribution function.

To generate a random variate X uniformly distributed in the interval (a,b), we use inverse transformation technique. Let $U = F(X) = \frac{X-a}{b-a}$, and we obtain

$$X = a + (b-a)U \qquad (2.20)$$

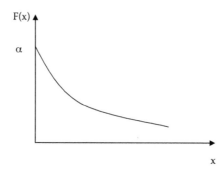

FIGURE 2.5
Exponential density function.

2.4.2 Exponential Distribution

This distribution, also known as the *negative exponential distribution*, is frequently used because of lack of conditioning of remaining time on past time expended. This peculiar characteristic is known as having Markov, *forgetfulness,* or *lack of memory* property. The distribution is portrayed in Figure. 2.5 and has the following characteristics:

$$f(x) = e^{-\alpha x}, \quad x > 0, \ \alpha > 0$$

$$F(x) = \int_a^x f(t)dt = 1 - e^{-\alpha x}, \quad x > 0$$

$$E(X) = \frac{1}{\alpha}$$

$$V(x) = \frac{1}{\alpha^2}$$

(2.21)

We use the inverse transform method to generate a random variate X that is exponentially distributed with mean μ. As

$$F(x) = 1 - e^{-x/\mu}, \quad 0 < x < \infty \tag{2.22}$$

Solving for X in $U = F(X)$ gives

$$X = -\mu \ln(1 - U) \tag{2.23}$$

Because $(1 - U)$ is itself a random number in the interval $(0,1)$, we simply write

$$X = -\mu \ln U \tag{2.24}$$

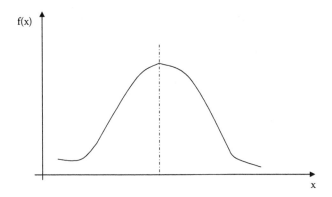

FIGURE 2.6
Normal or Gaussian density function.

2.4.3 Normal Distribution

This distribution, also known as the *Gaussian distribution*, is heavily used in probability and statistics, for reasons that will soon become obvious. It is used to describe phenomena with symmetric variations above and below the mean μ. The normal distribution has the form

$$f(x) = \frac{1}{\sigma\sqrt{2\pi}} \exp\left[-\frac{1}{2}\left(\frac{x-\mu}{\sigma}\right)^2\right], \quad -\infty < x < \infty \tag{2.25}$$

where the mean μ and the variance σ^2 are themselves incorporated in the pdf. Figure 2.6 shows the normal pdf. It is a convention to use the notation $X \sim N(\mu, \sigma^2)$ to denote a normal random variable X with mean μ and variance σ^2. When $\mu = 0$, $\sigma = 1$, we have the *standard normal* $N(0,1)$ distribution function with

$$f(z) = \frac{1}{\sqrt{2\pi}} e^{-z^2/2} \tag{2.26}$$

which is widely tabulated [12–14].

It is important to note the following points above the normal distribution, which make the distribution the most prominent in probability and statistics:

1. The binomial probability function with parameters m and p is approximated by a Gaussian pdf with $\mu = mp$ and $\sigma^2 = mp(1 - p)$ for large m and finite p.
2. The Poisson probability function with parameter γ can be approximated by a normal distribution with $\mu = \sigma^2 = \gamma$ for large γ.

3. The normal distribution is useful in characterizing the uncertainty associated with the estimated values. In other words, it is used in performing statistical analysis on simulation output, as we shall see in Section 2.5.

4. The justification for the use of the normal distribution comes from the *central limit theorem,* which, as will be discussed in Section 2.5, states that the distribution of the sum of n independent variables from any distribution approaches a normal distribution as n becomes large.

Thus, the normal distribution is used to model the cumulative effect of many small disturbances, each of which contributes to the stochastic variable X. It has the advantage of being mathematically tractable. Consequently, many statistical analyses such as those of regression and variance have been derived assuming a normal density function.

Because the function $f(z)$ is not invertible, the inverse transformation method cannot be used to generate random variate Z. However, several indirect methods have been used. For two independent standard normal variates z_1 and z_2 each with $N(0,1)$, the joint probability density function is

$$f(z_1, z_2) = \frac{1}{2\pi} \exp\left[-\frac{1}{2}(z_1^2 + z_2^3) \right] \tag{2.27}$$

By letting

$$z_1 = r\cos\theta \quad \text{and} \quad z_2 = r\sin\theta$$

we can generate two standard normal random variates Z_i and Z_{i+1} by taking two steps:

a. Generate two uniformly distributed independent random variates U_i and U_{i+1} defined over $(0,1)$.

b. Then obtain two independent and standard normally distributed variates Z_i and Z_{i+1} from

$$Z_i = (-2\ln U_i)^{1/2} \cos 2\pi U_{i+1}$$
$$Z_{i+1} = (-2\ln U_i)^{1/2} \sin 2\pi U_{i+1} \tag{2.28}$$

A random variable X from a normal distribution $N(\mu, \sigma^2)$ can be obtained from Z using

$$X = \sigma Z + \mu$$

This method is often recommended because it is easy to program.

An approximate popular method for generating normally distributed random variates is based on the central limit theorem. To generate a Gaussian variable X with mean μ and standard deviation σ, we follow these steps:

a. Generate 12 uniformly distributed random numbers $U_1, U_2, \cdots, U_{12}$.
b. Obtain $Z = \sum_{i=1}^{12} U_i - 6$.
c. Set $X = \sigma Z + \mu$.

This algorithm is used in most existing subroutines for generating Gaussian variables.

Other methods of generating random variates from exponential and normal distributions can be found in [22]. MATLAB® functions for generating uniform, exponential, and normal variates are shown in Figure 2.7.

```
function   x =uniform (a,b,m)

x= a + (b-a)*rand(m,1)
```

(a) **Uniform variates on interval (a,b).**

```
function   x = exponential (lambda, m)

x= - (1/lambda)*log(rand(m,1);
```

(b) **Exponential variates with mean 1/lambda.**

```
function x = gaussian(mu, sigma, m)

x = mu+ (sigma*randn(m,1))
```

(c) **Gaussian variates with mean mu and standard deviation sigma.**

FIGURE 2.7
Functions for generating random (a) uniform, (b) exponential, (c) Gaussian variates.

```
for k = 1:100

    r = rand;

    theta(k) = acos( 1 – 2*r);

end;
```

FIGURE 2.8
Random variable generator for Example 2.3.

Example 2.3

Using random number U uniformly distributed between 0 and 1, generate random variable Θ with probability distribution given by

$$T(\theta) = \frac{1}{2}(1 - \cos\theta), \quad 0 < \theta < \pi$$

Solution

We use the inverse transformation method. To generate the random variable Θ, set

$$U = T(\Theta) = \frac{1}{2}(1 - \cos\Theta),$$

then

$$\Theta = T^{-1}(U) = \cos^{-1}(1 - 2U)$$

Using this, a sequence of random variates Θ with the given distribution is generated in the program of Figure 2.8.

2.5   Evaluation of Error

Monte Carlo methods give solutions that are averages over a number of tests. For this reason, it is important to realize that the sample statistics obtained from Monte Carlo experiments will vary from one experiment to another. In fact, the sample statistics themselves are random variables and, as such, have associated probability distributions, means, variances, and standard deviation. Thus, the simulation results contain fluctuations about a mean value, and it is impossible to ascribe a 100% confidence in the results. To evaluate the statistical uncertainty or error in a simulation experiment, we must resort to various statistical techniques associated with random variables and utilize the central limit theorem.

Suppose that X is a random variable. Recall that we define the expected or mean value of X as

$$\mu = \int_{-\infty}^{\infty} x f(x)dx \tag{2.29}$$

where $f(x)$ is the probability density distribution of X. If we draw random and independent samples, $x_1, x_2, \cdots, x_N$ from $f(x)$, our estimate of x would take the form of the mean of N samples, namely,

$$\hat{\mu} = \frac{1}{N} \sum_{n=1}^{N} x_n \tag{2.30}$$

Although μ is the true mean value of X, $\hat{\mu}$ is the unbiased estimator of μ; an unbiased estimator being one with the correct expectation value. In general, $\hat{\mu} \neq \mu$. The standard deviation, defined as

$$\sigma(x) = (E(X^2) - \mu^2)^{1/2} \tag{2.31}$$

provides a measure of the spread in the values of $\hat{\mu}$ about μ; it yields the order of magnitude of the error. The confidence we place in the estimate of the mean is given by the variance of $\hat{\mu}$. The relationship between the variance of $\hat{\mu}$ and the variance of x is

$$\sigma(\hat{\mu}) = \frac{\sigma(x)}{\sqrt{N}} \tag{2.32}$$

This shows that if we use $\hat{\mu}$ constructed from N values of x_n according to Equation 2.30 to estimate μ, then the spread in our results of $\hat{\mu}$ about μ is proportional to $\sigma(x)$ and falls off as the number of samples N increases.

In order to estimate the spread in $\hat{\mu}$, we define the *sample variance*

$$S^2 = \frac{1}{N-1} \sum_{n=1}^{N} (x_n - \hat{\mu})^2 \tag{2.33}$$

Again, it can be shown that the expected value of S^2 is equal to $\sigma^2(x)$. Therefore, the sample variance is an unbiased estimator of $\sigma^2(x)$. Multiplying out the square term in Equation 2.33, it is readily shown that the *sample standard deviation*

$$S = \left(\frac{N}{N-1}\right)^{1/2} \left[\frac{1}{N} \sum_{n=1}^{N} x_n^2 - \hat{x}^2\right]^{1/2} \tag{2.34}$$

For large N, the factor $N/(N-1)$ is set equal to one.

As a way of arriving at the central limit theorem, a fundamental result in probability theory, consider the binomial function

$$B(M) = \frac{N!}{M!(N-M)!} p^M q^{N-M} \tag{2.35}$$

which is the probability of M successes in N independent trials. In Equation 2.35, p is the probability of success in a trial and $q = 1 - p$ is the probability of failure. If M and $N - M$ are large, we may use *Stirling's formula*

$$n! \sim n^n e^{-n} \sqrt{2\pi n} \tag{2.36}$$

so that Equation 2.35 is approximated as the normal distribution [23]:

$$B(M) \simeq f(\hat{x}) = \frac{1}{\sigma(\hat{x})\sqrt{2\pi}} \exp\left[-\frac{(\hat{x} - \bar{x})^2}{2\sigma^2(\hat{x})}\right] \tag{2.37}$$

where $\bar{x} = Np$ and $\sigma(\hat{x}) = \sqrt{Npq}$. Thus, as $N \to \infty$ the central limit theorem states that the probability density function which describes the distribution of $\hat{x}$ resulting from N simulation experiments is the normal distribution $f(\hat{x})$ in Equation 2.37. In other words, the sum of a large number of random variables tends to be normally distributed. Inserting Equation 2.32 into Equation 2.37 gives

$$f(\hat{x}) = \sqrt{\frac{N}{2\pi}} \frac{1}{\sigma(x)} \exp\left[-\frac{N(\hat{x} - \bar{x})^2}{2\sigma^2(\hat{x})}\right] \tag{2.38}$$

The normal (or Gaussian) distribution is very useful in various problems in engineering, physics, and statistics. The remarkable versatility of the Gaussian model stems from the central limit theorem. For this reason, the Gaussian model often applies to situations in which the quantity of interest results from the summation of many irregular and fluctuating components.

Because the number of samples N is finite, absolute certainty in simulation is unattainable. We try to estimate some limit or interval around μ such that we can predict with some confidence that $\hat{\mu}$ falls within that limit. Suppose we want the probability that $\hat{\mu}$ lies between $\mu - \varepsilon$ and $\mu + \varepsilon$. By definition,

$$\text{Prob}[\mu - \varepsilon < \hat{\mu} < \mu + \varepsilon] = \int_{\mu - \varepsilon}^{\mu + \varepsilon} f(\hat{\mu}) d\hat{\mu} \tag{2.39}$$

By letting $\lambda = \frac{(\hat{\mu} - \mu)}{\sqrt{2/N}\sigma(x)}$,

$$\text{Prob}[\mu - \varepsilon < \hat{\mu} < \mu + \varepsilon] = \frac{2}{\sqrt{\pi}} \int_0^{(\sqrt{N/2})(\varepsilon\sigma)} e^{-\lambda^2} d\lambda$$

$$= \text{erf}\left(\sqrt{N/2}\frac{\varepsilon}{\sigma(x)}\right) \tag{2.40}$$

or

$$\text{Prob}\left\{\mu - z_{\alpha/2}\frac{\sigma}{\sqrt{N}} \leq \hat{\mu} \leq \mu + z_{\alpha/2}\frac{\sigma}{\sqrt{N}}\right\} = 1 - \alpha \qquad (2.41)$$

where erf(x) is the error function and $z_{\alpha/2}$ is the upper $\alpha/2 \times 100$ percentile of the standard normal deviation. The random interval $\hat{x} \pm \varepsilon$ is called a *confidence interval*, and $\text{erf}(\sqrt{\frac{N}{2}}\frac{\varepsilon}{\sigma(x)})$ is the *confidence level*. Most simulation experiments use error $\varepsilon = \sigma(x)/\sqrt{N}$, which implies that $\hat{\mu}$ is within one standard deviation of μ, the true mean. From Equation 2.41, the probability that the sample mean $\hat{\mu}$ lies within the interval $\hat{\mu} \pm \sigma(x)/\sqrt{N}$ is 0.6826 or 68.3%. If higher confidence levels are desired, two or three standard deviations may be used. For example,

$$\text{Prob}\left(\mu - M\frac{\sigma(x)}{\sqrt{N}} < \hat{\mu} < \mu + M\frac{\sigma(x)}{\sqrt{N}}\right) = \begin{cases} 0.6826, & M = 1 \\ 0.954, & M = 2 \\ 0.997, & M = 3 \end{cases} \qquad (2.42)$$

where M is the number of standard deviations.

In Equations 2.41 and 2.42, it is assumed that the population standard deviation σ is known. Because this is rarely the case, σ must be estimated by the sample standard S calculated from Equation 2.34 so that the normal distribution is replaced by the Student's t-distribution. It is well known that the t-distribution approaches the normal distribution as N becomes large, say, $N > 30$. Thus, Equation 2.41 is equivalent to

$$\text{Prob}\left(\mu - \frac{St_{\alpha/2;N-1}}{\sqrt{N}} \leq \hat{\mu} \leq \mu + \frac{St_{\alpha/2;N-1}}{\sqrt{N}}\right) = 1 - \alpha \qquad (2.43)$$

where $t_{\alpha/2;N-1}$ is the upper $100 \times \alpha/2$ percentage point of Student's t-distribution with $(N - 1)$ degrees of freedom and its values are listed in any standard statistics text. The upper and lower limits of a confidence interval are given by

$$\text{upper limit} = \mu + \varepsilon \qquad (2.44)$$

$$\text{lower limit} = \mu - \varepsilon \qquad (2.45)$$

where

$$\varepsilon = \frac{St_{\alpha/2;N-1}}{\sqrt{N}} \qquad (2.46)$$

Thus, if a simulation is performed N times (by using different seed values), then in $(1 - \alpha)$ of cases, the estimate $\hat{\mu}$ lies within the confidence interval,

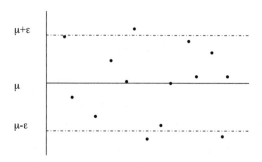

FIGURE 2.9
Confidence of interval.

whereas in α cases the estimate lies outside the interval, as illustrated in Figure 2.9. Equation 2.46 provides the error estimate for a given number, N, of simulation experiments or observations.

If, on the other hand, an accuracy criterion ε is prescribed and we want to estimate μ by $\hat{\mu}$ within tolerance of ε with at least probability $1-\alpha$, we must ensure that the sample size N satisfies

$$\text{Prob}[\,|\hat{\mu}-\mu|<\varepsilon\,]\geq 1-\alpha \tag{2.47}$$

To satisfy this requirement, N must be selected as a small integer satisfying

$$N \geq \left(\frac{St_{\alpha/2;N-1}}{\sqrt{\varepsilon}}\right)^2 \tag{2.48}$$

For further discussion on error estimates in simulation, one should consult [24,25].

Example 4.4

In a simulation experiment, an analyst obtained the mean values of a certain parameter as 7.60, 6.60, 6.97, 7.50, and 7.43 for five simulation runs using different seed values. Calculate the error estimate using 95% confidence interval.

Solution

We first get the sample mean

$$\mu = \frac{(7.60+6.60+6.97+7.50+7.43)}{5} = 7.22$$

From Equation 2.33, the sample variance is obtained as

$$S^2 = \frac{(7.60-7.22)^2 + \cdots + (7.43-7.22)^2}{4} = 0.23793$$

or $S = 0.48778$.

Using 95% confidence interval, $1 - \alpha = 95\%$, that is, $\alpha = 0.05$. For five runs ($N = 5$), the t-distribution table gives $t_{\alpha/2;N-1} = 2.776$. Using Equation 2.46, the error is estimated as

$$\varepsilon = \frac{0.48778 \times 2.776}{5} = 0.6056$$

Thus, the 95% confidence interval for the parameter is

$$\mu - \varepsilon < \hat{\mu} < \mu + \varepsilon = 6.6144 < \hat{\mu} < 7.8265$$

2.6 Summary

This chapter briefly reviews some fundamental concepts in probability and statistics relating to Monte Carlo simulation of EM problems. Although procedures for generating pseudorandom numbers are discussed, one need not generate uniform random numbers in most cases. Random number generators are usually built into a simulation package or available in a digital computer. However, if one must develop a random number generator, testing it before application is essential to ensure a large measure of confidence in the simulation results. The chapter also discusses how random variates are generated from probability distributions. Additional information on these concepts can be obtained from texts on probability and statistics [7,8,10,21].

Using random numbers always provides simulation output that is subject to random error. Therefore, one would like to know how close is the point estimate to the mean value μ it is supposed to estimate. The statistical accuracy of point estimates is measured in terms of the confidence interval. By generating a number of observations or simulation experiments, say N, and employing a standard statistical method, the error estimate can be obtained.

References

1. G. S. Fishman, *Concepts and Methods in Discrete Event Digital Simulation*, New York: John Wiley & Sons, 1973, pp. 167–241.
2. T. E. Hull and A. R. Dobell, Random number generators, *SIAM Review*, vol. 4, no. 3, 230–254, July 1982.
3. D. E. Knuth, *The Art of Computer Programming*. Reading, MA: Addison-Wesley, vol. 2, 1969, pp. 9,10,78,155.

4. J. Banks and J. Carson, *Discrete-Event System Simulation*. Englewood Cliffs, NJ: Prentice-Hall, 1984, pp. 256–329.

5. P. A. W. Lewis et al., A Pseudo-random Number Generator for the System/360, *IBM System J.*, vol. 8, no. 2, 136–146, 1969.

6. A. M. Law *and* W. D. Kelton, *Simulation Modeling and Analysis*, 2nd ed., New York: McGraw-Hill, 1991, pp. 462–521.

7. F. Neelamkavil, *Computer Simulation and Modelling*, Chichester: U.K., John Wiley & Sons, 1987, pp. 105–134.

8. R. Y. Rubinstein, *Simulation and the Monte Carlo Method*, New York: John Wiley & Sons, 1981, pp. 20–113.

9. J. P. C. Kleijnen and B. Annink, Pseudorandom number generators for super-computers and classical computers: A practical introduction, *Eur. J. Operational Res.*, vol. 63, no. 1, 76–85, 1992.

10. U. W. Pooch and J. A. Wall, *Discrete Event Simulation: A Practical Approach*. Boca Raton: CRC Press, 1993, pp. 145–175, chap. 7.

11. L. H. Sim and K. N. Nitschke, Testing random number generators for Monte Carlo applications, *Australasian Phys. Eng. Sci. Med.*, vol. 16, no. 1, 1993, pp. 22–32.

12. H. D. Larsen, *Rinehart Mathematical Tables, Formulas, and Curves*, New York: Rinehart, 1948, pp. 154–155.

13. R. S. Burington, *Handbook of Mathematical Tables and Formulas*, New York: McGraw-Hill, 1972, pp. 424–427.

14. W. M. Meredith, *Basic Mathematical and Statistical Tables for Psychology and Education*, New York: McGraw-Hill, 1969, pp. 186–189.

15. W. Bruns et al., *Monte Carlo Applications in Polymer Science*, Berlin/New York: Springer-Verlag, 1981, pp. 1–22.

16. J. R. Emshoff and R. L. Sisson, *Design and Use of Computer Simulation Models*, New York: Macmillan, 1970, pp. 170–182.

17. H. Kobayashi, *Modeling and Analysis: An Introduction to System Performance Evaluation Methodology*, Reading, MA: Addison-Wesley, 1978, pp. 221–247.

18. I. M. Sobol, *The Monte Carlo Method*, Chicago, IL: University of Chicago Press, 1974, pp. 24–30.

19. Y. A. Shreider, *Method of Statistical Testing (Monte Carlo Method)*, Amsterdam: Elsevier, 1964, pp. 39–83. Another translation of the same Russian text: Y. A. Shreider, *The Monte Carlo Method (The Method of Statistical Trials)*. Oxford: Pergamon, 1966.

20. H. A. Taha, *Simulation Modeling and SIMNET*, Englewood Cliffs, NJ: Prentice-Hall, 1988, pp. 45–52.

21. T. H. Naylor et al., *Computer Simulation Techniques*, New York: John Wiley & Sons, 1966, p. 108.

22. J. H. Ahrens and U. Deiter, Computer methods for sampling from the exponential and normal distributions, *Comm. ACM*, vol. 15, no. 10, 873–882, October 1972.

23. I. S. Sokolinkoff and R. M. Redheffer, *Mathematics of Physics and Modern Engineering*, New York: McGraw-Hill, 1958, pp. 644–649.

24. M. H. Merel and F. J. Mullin, Analytic Monte Carlo error analysis, *J. Spacecraft*, vol. 5, no. 11, 1304–1308, November 1968.

25. A. J. Chorin, Hermite expansions in Monte-Carlo computation, *J. Comp. Phys.*, vol. 8, 472–482, 1971.

Problems

2.1a. Consider the multiplicative congruential generator with $a = 7$, $m = 1$, and $Z_0 = 1$. Generate the first six numbers.

  b. Repeat Part a for $a = 11$, $m = 128$, and $Z_0 = 3$.

2.2 Beginning with $Z_0 = 34$, generate 10 pseudorandom numbers using a mixed congruential generator with $a = 12$, $c = 2$, and $m = 100$.

2.3a. Given $Z_{i+1} = 6Z_i$ (mod 11), $Z_0 = 1$, determine the period.

  b. Repeat (a) for $Z_{i+1} = 11Z_i$ (mod 18), $Z_0 = 5$.

2.4. Use the generator

$$Z_{i+1} = 121Z_i + 567 \quad (\text{mod } 1000), \quad Z_0 = 1$$

to produce 1000 uniformly distributed random numbers in the interval (0,1). Use the moment tests to check whether the generator is satisfactory.

2.5a. Write a program to generate 2000 random numbers uniformly distributed in the interval (0,1).

  b. Using the built-in function in your computing system, generate a set of 2000 random numbers.

  c. Divide the interval equally into 20 subintervals. Apply the frequency test, and compare the output of the two generators.

2.6a. Write a program to generate 5000 uniformly distributed random numbers in the interval (0,1) using the multiplicative congruential generator

$$Z_{i+1} = 101Z_i + 1 \quad (\text{mod } 1000), \quad Z_0 = 98$$

  b. By dividing the unit interval into 10 equal parts, use the frequency test to see whether the generator is a good one.

2.7. Generate random variates from $f(x) = 6x^2$, $0 \le x \le 1$ by: (a) the inverse transform and (b) the rejection method.

2.8. Uniformly distributed random integers between 11 and 30, inclusive, are to be generated from the random numbers U in the following table. How many of the integers are odd numbers?

0.04493 0.88159 0.52494 0.96119

0.75248 0.63896 0.33824 0.54692

0.45862 0.82391 0.51025 0.23287

0.61962 0.29529 0.79335 0.35963

0.65337 0.15307 0.12472 0.26898

2.9a. Generate 500 random variates, exponentially distributed with mean 6.

   b. Compute the sample mean, variance, and standard deviation of the variate.

2.10. A random sample of 50 variables taken from a normal population has a mean of 20 and standard deviation 8. Calculate the error with 95% confidence limits.

2.11. In a simulation model of a potential system, an analyst obtained the mean potentials for four simulation runs as 42.80, 41.60, 42.48, and 41.80 V. Calculate the 98% confidence interval for the potential.

3

Finite Difference Method

A point of view can be dangerous luxury when substituted for insight
and understanding.

—**Marshall McLuhan**

To apply Monte Carlo methods (MCMs) to solve the partial differential equations (PDEs) encountered in electromagnetics, we must first convert the differential equations into their finite difference equivalents. This requires familiarity with the finite difference method (FDM) and provides justification for including this chapter in this book.

The FDM was first developed by A. Thom in 1920 under the name "the method of squares" to solve nonlinear hydrodynamic equations. Since then the method has been used for solving problems occurring in various engineering disciplines, including electromagnetics. Section 3.1 offers the general approach to converting a PDE to its finite difference equivalent. We apply the scheme in Sections 3.2, 3.3, and 3.4 for parabolic, hyperbolic, and elliptic PDEs, respectively. Section 3.5 deals with stability and accuracy. In Section 3.6, we finally apply finite differences to solve Maxwell's equations.

3.1 Finite Differences

The first step in solving an electromagnetic boundary value problem using finite differences is to obtain difference equations. Given a function $f(x)$ shown in Figure 3.1, one can approximate the derivative, the slope of the tangent drawn at P, using the values at $A(x_o - h, f(x_o - h))$, $P(x_o, f(x_o))$, and $B(x_o + h, f(x_o + h))$ in one of three possible ways [1,2]:

$$f'(x) \approx \frac{\Delta f}{h} = \frac{f(x+h) - f(x)}{h} \qquad (3.1)$$

$$f'(x) \approx \frac{\nabla f}{h} = \frac{f(x) - f(x-h)}{h} \qquad (3.2)$$

$$f'(x) \approx \frac{\delta f}{h} = \frac{f(x+h/2) - f(x-h/2)}{h} \qquad (3.3)$$

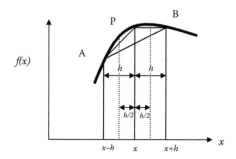

FIGURE 3.1
Approximating the derivative of *f(x)* at P using forward, backward, and central differences.

The symbols Δ, ∇, and δ respectively represent forward, backward (in the present context), and central difference operators. From Figure 3.1, we understand that the slope of the tangent at P is approximately equal to that of the chords PB, AB (approximately), and PA. Taylor's series expansion of *f(x)* around A and B reveals that, the forward, backward, and central differences are of orders h, h, and h^2, respectively. Thus, as the step size h becomes smaller and smaller, we expect truncation error to reduce, but the round-off errors occurring in the calculations grow bigger and bigger. This is suggested in Figure 3.2. So, one has to make a trade-off in the choice of step size.

Using the central difference, the second derivative at P may be expressed as

$$f''(x) \approx \frac{\delta f'}{h} = \frac{f(x+h) - 2f(x) + f(x-h)}{h^2} \tag{3.4}$$

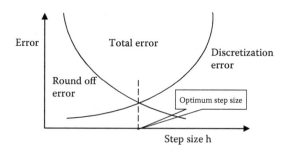

FIGURE 3.2
Trade-off between discretization error and round-off error.

The approach used here in obtaining finite difference approximations is rather intuitive. A more general approach is to use Taylor's series. Consider the following well-known expansions:

$$f(x+h) = f(x) + hf'(x) + \frac{1}{2!}h^2 f''(x) + \frac{1}{3!}h^3 f'''(x) + \cdots \qquad (3.5)$$

and

$$f(x-h) = f(x) - hf'(x) + \frac{1}{2!}h^2 f''(x) - \frac{1}{3!}h^3 f'''(x) + \cdots \qquad (3.6)$$

By adding these expansions, we get

$$f(x+h) + f(x-h) = 2f(x) + h^2 f''(x) + O(h^4) \qquad (3.7)$$

where $O(h^4)$ is the error introduced by truncating the series. It represents terms that are not greater than h^4. Assuming that these terms are negligible,

$$f''(x) \simeq \frac{f(x+h) - 2f(x) + f(x-h)}{h^2}$$

which is Equation 3.4. Subtracting Equation 3.6 from Equation 3.5 and neglecting terms of order h^3 yields

$$f'(x) \simeq \frac{f(x+h) - f(x-h)}{2h}$$

which is similar to Equation 3.3. This shows that the leading errors in Equations 3.3 and 3.4 are of the order h^2. Similarly, the difference formula in Equations 3.1 and 3.2 have truncation errors of the order $O(h)$. Higher-order finite difference approximations can be obtained by taking more terms in Taylor series expansion. If the infinite Taylor series was retained, an exact solution would be realized for the problem. However, for practical reasons, the infinite series is usually truncated after the second-order term. This imposes an error, which exists in all finite different solutions.

To apply the difference method to the solution of a PDE involving a function $\Phi(x,t)$, we divide the solution region in the $x-t$ plane into equal rectangles or meshes of sides Δx and Δt as shown in Figure 3.3. We let the coordinates (x,t) of a typical grid point or node be

$$x = i\Delta x, \quad i = 0, 1, 2, \ldots$$
$$t = j\Delta t, \quad j = 0, 1, 2, \ldots \qquad (3.8a)$$

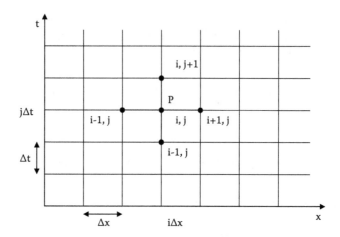

FIGURE 3.3
Finite difference mesh for two independent variables x and t.

and the value of Φ at P be

$$\Phi_p = \Phi(i\Delta x, j\Delta t) = \Phi(i, j) \tag{3.8b}$$

With this notation, the central difference approximations of the derivatives of Φ at the (i, j)-th node are

$$\Phi_x(i, j) \simeq \frac{\Phi(i+1, j) - \Phi(i-1, j)}{2\Delta x} \tag{3.9a}$$

$$\Phi_t(i, j) \simeq \frac{\Phi(i, j+1) - \Phi(i, j-1)}{2\Delta t} \tag{3.9b}$$

$$\Phi_{xx}(i, j) \simeq \frac{\Phi(i+1, j) - 2\Phi(i, j) + \Phi(i-1, j)}{(\Delta x)^2} \tag{3.9c}$$

$$\Phi_{tt}(i, j) \simeq \frac{\Phi(i, j+1) - 2\Phi(i, j) + \Phi(i, j-1)}{(\Delta t)^2} \tag{3.9d}$$

Table 3.1 provides some useful finite difference approximations for Φ_x and Φ_{xx}.

3.2 Finite Differencing of Parabolic PDEs

Consider a simple example of a parabolic (or diffusion) PDE with one spatial independent variable:

$$k\frac{\partial \Phi}{\partial t} = \frac{\partial^2 \Phi}{\partial x^2} \tag{3.10}$$

TABLE 3.1

Finite Difference Approximations for Φ_x and Φ_{xx}

Derivative	Finite Difference Approximation	Type	Error
Φ_x	$\dfrac{\Phi_{i+1} - \Phi_i}{h}$	FD	$O(h)$
	$\dfrac{\Phi_i - \Phi_{i-1}}{h}$	BD	$O(h)$
	$\dfrac{\Phi_{i+1} - \Phi_{i-1}}{h}$	CD	$O(h^2)$
	$\dfrac{-\Phi_{i+2} + 4\Phi_{i+1} - 3\Phi_i}{2h}$	FD	$O(h^2)$
	$\dfrac{3\Phi_i - 4\Phi_{i-1} + \Phi_{i-2}}{2h}$	BD	$O(h^2)$
	$\dfrac{-\Phi_{i+2} + 8\Phi_{i+1} - 8\Phi_{i-1} + \Phi_i}{12h}$	CD	$O(h^4)$
Φ_{xx}	$\dfrac{\Phi_{i+2} - 2\Phi_{i+1} + \Phi_i}{h^2}$	FD	$O(h^2)$
	$\dfrac{\Phi_{i+2} - 2\Phi_{i+1} + \Phi_i}{h^2}$	BD	$O(h^2)$
	$\dfrac{\Phi_{i+1} - 2\Phi_i + \Phi_{i-1}}{h^2}$	CD	$O(h^2)$
	$\dfrac{-\Phi_{i+2} + 16\Phi_{i+1} - 30\Phi_i + 16\Phi_{i-1} - \Phi_{i-2}}{12h^2}$	CD	$O(h^4)$

Note: $h = \Delta x$, FD = forward difference, BD = backward difference, and CD = central difference.

where k is constant. The equivalent finite difference approximation is

$$k\frac{\Phi(i, j+1) - \Phi(i, j)}{\Delta t} = \frac{\Phi(i+1, j) - 2\Phi(i, j) + \Phi(i-1, j)}{(\Delta x)^2} \tag{3.11}$$

where $x = i\Delta x$, $i = 0, 1, 2, \ldots n$, $t = j\Delta t$, $j = 0, 1, 2, \ldots$. In Equation 3.11, we have used the forward difference formula for the derivative with respect to t and central difference formula for that with respect to x. If we let

$$r = \frac{\Delta t}{k(\Delta x)^2} \tag{3.12}$$

Equation 3.11 can be written as

$$\Phi(i, j+1) = r\Phi(i+1, j) + (1 - 2r)\Phi(i, j) + r\Phi(i-1, j) \tag{3.13}$$

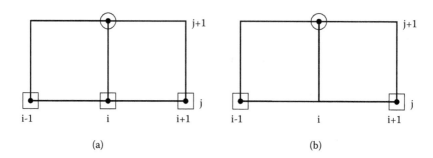

FIGURE 3.4
Computational molecule for parabolic PDE: (a) for $0 < r \le \frac{1}{2}$, (b) for $r = 1/2$.

This *explicit formula* can be used to compute $\Phi(x, t + \Delta t)$ explicitly in terms of $\Phi(x,t)$. Thus, the values of Φ along the first time row (see Figure 3.3), $t = \Delta t$, can be calculated in terms of the boundary and initial conditions, then the values of Φ along the second time row, $t = 2\Delta t$, are calculated in terms of the first time row, and so on.

A graphical way of describing the difference formula of Equation 3.13 is through the computational molecule of Figure 3.4a, where the square is used to represent the grid point where Φ is presumed to be known and a circle where Φ is unknown.

To ensure a stable solution or reduce errors, care must be exercised in selecting the value of r in Equations 3.12 and 3.13. It will be shown in Section 3.5 that Equation 3.13 is valid only if the coefficient $(1 - 2r)$ in Equation 3.13 is nonnegative or $0 < r \le \frac{1}{2}$. If we choose $r = \frac{1}{2}$, Equation 3.13 becomes

$$\Phi(i, j+1) = \frac{1}{2}[\Phi(i+1, j) + \Phi(i-1, j)] \tag{3.14}$$

so that the computational molecule becomes that shown in Figure 3.4b.

The fact that obtaining stable solutions depends on r or the size of the time step Δt renders the explicit formula in Equation 3.13 inefficient. Although the formula is simple to implement, its computation is slow. An *implicit formula*, proposed by Crank and Nicholson in 1974, is valid for all values of r. We replace $\partial^2\Phi/\partial x^2$ in Equation 3.10 by the average of the central difference formulas on the j-th and $(j + 1)$-th time rows so that

$$k\frac{\Phi(i, j+1) - \Phi(i, j)}{\Delta t} = \frac{1}{2}\left[\frac{\Phi(i+1, j) - 2\Phi(i, j) + \Phi(i-1, j)}{(\Delta x)^2}\right.$$
$$\left. + \frac{\Phi(i+1, j+1) - 2\Phi(i, j+1) + \Phi(i-1, j+1)}{(\Delta x)^2}\right]$$

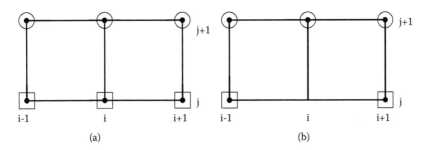

FIGURE 3.5
Computational molecule for Crank–Nicholson method: (a) for finite values of r, (b) for $r = 1$.

This can be rewritten as

$$-r\Phi(i-1, j+1) + 2(1+r)\Phi(i, j+1) - r\Phi(i+1, j+1) = r\Phi(i-1, j) + 2(1-r)\Phi(i, j)$$

$$+r\Phi(i+1, j) \qquad (3.15)$$

where r is given by Equation 3.12. The right-hand side of Equation 3.15 consists of three known values, whereas the left-hand side has the three unknown values of Φ. This is illustrated in the computational molecule of Figure 3.5a. Thus, if there are n free nodes along each time row, then for $j = 0$, applying Equation 3.15 to nodes $i = 1, 2, \ldots, n$ results in n simultaneous equations with n unknown values of Φ and known initial and boundary values of Φ. Similarly, for $j = 1$, we obtain n simultaneous equations for n unknown values of Φ in terms of the known values for $j = 0$, and so on. The combination of accuracy and unconditional stability allows the use of a much larger time step with Crank–Nicholson method than is possible with the explicit formula. Although the method is valid for all finite values of r, a convenient choice of $r = 1$ reduces Equation 3.15 to

$$-\Phi(i-1, j+1) + 4\Phi(i, j+1) - \Phi(i+1, j+1) = \Phi(i-1, j) + \Phi(i+1, j) \quad (3.16)$$

with the computational molecule of Figure 3.5b.

More complex finite difference schemes can be developed by applying the same principles discussed here. Two such schemes are the leapfrog method and the Dufort–Frankel method [3,4]. These schemes, as well as those discussed earlier, are summarized in Table 3.2. Note that the last two methods are two-step finite difference schemes in finding Φ at time $j + 1$ that require knowing Φ at two previous time steps j and $j - 1$, whereas the first two methods are one-step schemes.

TABLE 3.2

Finite Difference Approximation to the Parabolic Equation: $\dfrac{\partial \Phi}{\partial t} = \dfrac{1}{k}\dfrac{\partial^2 \Phi}{\partial x^2}$

Method	Algorithm	Molecule
1. First order (Euler)	$\dfrac{\Phi_i^{j+1}-\Phi_i^{j}}{\Delta t} = \dfrac{\Phi_{i+1}^{j}-2\Phi_i^{j}+\Phi_{i-1}^{j}}{k(\Delta x)^2}$ explicit, stable for $r = \Delta t/k(\Delta x)^2 \le 0.5$	
2. Crank–Nicholson	$\dfrac{\Phi_i^{j+1}-\Phi_i^{j}}{\Delta t} = \dfrac{\Phi_{i+1}^{j+1}-2\Phi_i^{j+1}+\Phi_{i-1}^{j+1}}{2k(\Delta x)^2}$ $+\dfrac{\Phi_{i+1}^{j}-2\Phi_i^{j}+\Phi_{i-1}^{j}}{2k(\Delta x)^2}$ implicit, always stable	
3. Leapfrog	$\dfrac{\Phi_i^{j+1}-\Phi_i^{j-1}}{2\Delta t} = \dfrac{\Phi_{i+1}^{j}-2\Phi_i^{j}+\Phi_{i-1}^{j}}{k(\Delta x)^2}$ explicit, always unstable	
4. Dufort–Frankel	$\dfrac{\Phi_i^{j+1}-\Phi_i^{j-1}}{2\Delta t} = \dfrac{\Phi_{i+1}^{j}-\Phi_i^{j+1}-\Phi_i^{j-1}+\Phi_{i-1}^{j}}{k(\Delta x)^2}$ explicit, unconditionally stable	

Example 3.1

Solve the diffusion equation

$$\frac{\partial^2 \Phi}{\partial x^2} = \frac{\partial \Phi}{\partial t}, \quad 0 \le x \le 1 \tag{3.17}$$

subject to the boundary conditions

$$\Phi(0,t) = 0 = \Phi(1,t) = 0, \quad t > 0 \tag{3.18a}$$

and initial condition

$$\Phi(x, 0) = 100 \tag{3.18b}$$

Solution

This problem may be regarded as a mathematical model of the temperature distribution in a rod of length $L = 1$ m with its ends in contact with ice blocks (or held at 0°C), the rod being initially at 100°C. With this physical interpretation, our problem is finding the internal temperature Φ as a function of position and time. We will solve this problem using both explicit and implicit methods.

(a) Explicit Method

For easy hand calculation, let us choose $\Delta x = 0.1$, $r = \frac{1}{2}$ so that

$$\Delta t = kr(\Delta x)^2 = 0.005$$

as $k = 1$. We need the solution for only $0 \le x \le 0.5$ because the problem is symmetric with respect to $x = 0.5$. First, we calculate the initial and boundary values using Equation 3.18. These values of Φ at the fixed nodes are shown in Table 3.3 for $x = 0$, $x = 1$, and $t = 0$. Note that the values of $\Phi(0,0)$ and $\Phi(1,0)$ are taken as the average of 0 and 100. We now calculate Φ at the free nodes using Equation 3.14 or the molecule of Figure 3.4b. The result is shown in Table 3.3. The analytic solution to Equation 3.17 subject to Equation 3.18 is

$$\Phi(x,t) = \frac{400}{\pi} \sum_{m=0}^{\infty} \frac{1}{n} \sin n\pi x \exp(-n^2\pi^2 t), \quad n = 2m+1$$

Comparison of the explicit finite difference solution with the analytic solution at $x = 0.4$ is shown in Table 3.4. The table shows that the finite difference solution is reasonably accurate. Greater accuracy can be achieved by choosing smaller values of Δx and Δt.

(b) Implicit Method

Let us choose $\Delta x = 0.2$, $r = 1$ so that $\Delta t = 0.04$. The values of Φ at the fixed nodes are calculated as in part (a) of the example (see Table 3.3). For the free

TABLE 3.3

Result of Example 3.1

t x	0	0.1	0.2	0.3	0.4	0.5	0.6	...	1.0
0	50	100	100	100	100	100	100		50
0.005	0	75.0	100	100	100	100	100		0
0.01	0	50	87.5	100	100	100	100		0
0.015	0	43.75	75	93.75	100	100	100		0
0.02	0	37.5	68.75	87.5	96.87	100	96.87		0
0.025	0	24.37	62.5	82.81	93.75	96.87	93.75		0
0.03	0	31.25	58.59	78.21	89.84	93.75	89.84		0
⋮									
0.1	0	14.66	27.94	38.39	45.18	47.44	45.18		0

TABLE 3.4

Comparison of Explicit Finite Difference Solution with Analytic
Solution for Example 3.1

t	Finite Difference Solution at $x = 0.4$	Analytic Solution at $x = 0.4$	Percentage Error
0.005	100	99.99	0.01
0.01	100	99.53	0.47
0.015	100	97.85	2.2
0.02	96.87	95.18	1.8
0.025	93.75	91.91	2.0
0.03	89.84	88.32	1.7
0.035	85.94	84.61	1.6
0.04	82.03	80.88	1.6
⋮			
0.10	45.18	45.13	0.11

nodes, we apply Equation 3.16 or the molecule of Figure 3.5b. If we denote
$\Phi(i, j + 1)$ by Φ_i ($i = 1, 2, 3, 4$), the values of Φ for the first time step (Figure 3.6)
can be obtained by solving the following simultaneous equations:

$$-0 + 4\Phi_1 - \Phi_2 = 50 + 100$$

$$-\Phi_1 + 4\Phi_2 + \Phi_3 = 100 + 100$$

$$-\Phi_2 + 4\Phi_3 - \Phi_4 = 100 + 100$$

$$-\Phi_3 + 4\Phi_4 - 0 = 100 + 50$$

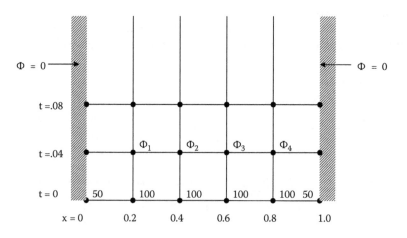

FIGURE 3.6
Figure for Example 3.1, part (b).

Solving these equations, we get

$$\Phi_1 = 58.13, \quad \Phi_2 = 82.54, \quad \Phi_3 = 72, \quad \Phi_4 = 55.5$$

at $t = 0.04$. Using the values of Φ, we apply Equation 3.16 to obtain another set of simultaneous equations for $t = 0.08$ as

$$-0 + 4\Phi_1 - \Phi_2 = 0 + 82.5$$

$$-\Phi_1 + 4\Phi_2 + \Phi_3 = 58.13 + 72$$

$$-\Phi_2 + 4\Phi_3 - \Phi_4 = 82.54 + 55.5$$

$$-\Phi_3 + 4\Phi_4 - 0 = 72 + 0$$

which results in

$$\Phi_1 = 34.44, \quad \Phi_2 = 55.23, \quad \Phi_3 = 56.33, \quad \Phi_4 = 32.08$$

This procedure can be programmed and accuracy can be increased by checking more points for each time step.

3.3 Finite Differencing of Hyperbolic PDEs

The simplest hyperbolic partial differential equation is the wave equation of the form

$$u^2 \frac{\partial^2 \Phi}{\partial x^2} = \frac{\partial^2 \Phi}{\partial t^2} \tag{3.19}$$

where u is the speed of the wave. An equivalent finite difference formula is

$$u^2 \frac{\Phi(i+1, j) - 2\Phi(i, j) + \Phi(i-1, j)}{(\Delta x)^2} = \frac{\Phi(i, j+1) - 2\Phi(i, j) + \Phi(i, j-1)}{(\Delta t)^2}$$

where $x = i\Delta x$, $t = j\Delta t$, and $i,j = 0, 1, 2, \ldots$. This equation can be written as

$$\Phi(i, j+1) = 2(1-r)\Phi(i, j) + r[\Phi(i+1, j) + \Phi(i-1, j)] - \Phi(i, j-1) \tag{3.20}$$

where $\Phi(i,j)$ is an approximation to $\Phi(x,t)$ and r is the *aspect ratio* given by

$$r = \left(\frac{u\Delta t}{\Delta x} \right)^2 \tag{3.21}$$

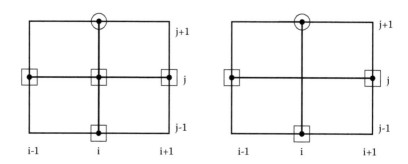

FIGURE 3.7
Computational molecule for wave equation: (a) for arbitrary $r \le 1$, (b) for $r = 1$.

Equation 3.20 is an explicit formula for the wave equation. The corresponding computational molecule is shown in Figure 3.7a. For the solution in Equation 3.20 to be stable, the aspect ratio $r \le 1$, as will be shown in Example 3.5. If we choose $r = 1$, Equation 3.20 becomes

$$\Phi(i, j+1) = \Phi(i+1, j) + \Phi(i-1, j) - \Phi(i, j-1) \qquad (3.22)$$

with the computational molecule in Figure 3.7b. Unlike the single-step schemes of Equations 3.13 and 3.15, the two-step schemes of Equations 3.20 and 3.22 require that the values of Φ at times j and $j - 1$ be known to get Φ at time $j + 1$. Thus, we must derive a separate algorithm to "start" the solution of Equation 3.20 or 3.22; that is, we must compute $\Phi(i, 1)$ and $\Phi(i, 2)$. To do this, we utilize the prescribed initial condition. For example, suppose the initial condition on the PDE in Equation 3.19 is

$$\left.\frac{\partial \Phi}{\partial t}\right|_{t=0} = 0$$

We use the backward difference formula

$$\frac{\partial \Phi(x, 0)}{\partial t} \simeq \frac{\Phi(i, 1) - \Phi(i, -1)}{2\Delta t} = 0$$

or

$$\Phi(i, 1) = \Phi(i, -1) \qquad (3.23)$$

Substituting Equation 3.23 into Equation 3.20 and taking $j = 0$ (i.e., at $t = 0$), we get

$$\Phi(i, 1) = 2(1 - r)\Phi(i, 0) + r[\Phi(i+1, 0) + \Phi(i-1, 0)] - \Phi(i, -1)$$

or

$$\Phi(i, 1) = (1 - r)\Phi(i, 0) + \frac{r}{2}[\Phi(i+1, 0) + \Phi(i-1, 0)] \qquad (3.24)$$

Using the starting formula in Equation 3.24 together with the prescribed boundary and initial conditions, the value of Φ at any point (i,j) can be obtained directly from Equation 3.20.

There are implicit methods for solving hyperbolic PDEs just as we have implicit methods for parabolic PDEs. However, for hyperbolic PDEs, implicit methods require an infinite number of simultaneous equations to be solved and therefore cannot be used without making some simplifying assumptions.

Example 3.2

Solve the wave equation

$$\Phi_{tt} = \Phi_{xx}, \quad 0 < x < 1, \ t \geq 0$$

subject to the boundary conditions

$$\Phi(0,t) = 0 = \Phi(1,t), \quad t \geq 0$$

and initial conditions

$$\Phi(x,0) = \sin \pi x, \quad 0 < x < 1$$

$$\Phi_t(x,0) = 0, \qquad 0 < x < 1$$

Solution

The analytic solution is easily obtained as

$$\Phi(x,t) = \sin \pi x \cos \pi t \tag{3.25}$$

Using the explicit finite difference of Equation 3.20 with $r = 1$, we obtain

$$\Phi(i, j+1) = \Phi(i-1, j) + \Phi(i+1, j) - \Phi(i, j-1), \quad j \geq 1 \tag{3.26}$$

For $j = 0$, substituting

$$\Phi_t = \frac{\Phi(i,1) - \Phi(i,-1)}{2\Delta t} = 0$$

or

$$\Phi(i, 1) = \Phi(i, -1)$$

into Equation 3.20 gives the following starting formula:

$$\Phi(i,1) = \frac{1}{2}[\Phi(i-1,0) + \Phi(i+1,0)] \tag{3.27}$$

as $u = 1$, and $r = 1$, $\Delta t = \Delta x$. Also, because the problem is symmetric with respect to $x = 0.5$, we solve for Φ using Equations 3.26 and 3.27 within $0 < x < 0.5$,

```
% ********************************************************************************
% MATLAB code for Example 3.2 on one-dimensional wave equation solved
% using an explicit finite difference scheme
% ******************************************************************************** **

clear all; format compact; tic

% Explicit Method
    delx = 0.1;        % resolution size
    r = 1;             % 'aspect ratio'
    u = 1;             % Constant of given wave equation
    delt = r^2*delx/u; % time step size
    Tsteps = round(1/delt); % Number of time steps

    % X1 is the potential grid of the simulation, due to symetry only half
    % of the field is calculated.
    X1 = zeros(Tsteps,1/(2*delx)+2);    % Initilize X1

    % Initial conditions and reflection line defined
    x = 0:delx:.5+delx;
    X1(1,:) = sin(pi*x);
    X1(2,2:end-1) = .5*( X1(1,1:end-2)+X1(1,3:end));
    X1(2,end) = X1(2,end-2); % reflection line

    for row = 3:size(X1,1)
        for col = 2:size(X1,2)-1
            X1(row,col) = X1(row-1,col-1)+X1(row-1,col+1)-X1(row-2,col); % eqn. (3.26)
        end
        X1(row,end) = X1(row,end-2);    % reflected line
    end

    % Use symetry condition to create entire field
    X2 = [X1,fliplr(X1(:,1:end-3))];

    figure(1),imagesc(0:delx:1,(0:delt:Tsteps*delt),X2),colorbar
        ylabel('\leftarrow time (sec)')
        xlabel('x')
        title('Hyperbolic PDE')

    if (delx==.1)
        dispmat = [X1(1:8,1:7)];
        disp(sprintf('\nCompare to Table 3.5, Solution of the Wave Equation in Example
3.2'))
        disp(num2str(dispmat))
    end
```

FIGURE 3.8
MATLAB code for Example 3.2.

TABLE 3.5

Solution of the Wave Equation in Example 3.2

xt	0	0.1	0.2	0.3	0.4	0.5	0.6
0.0	0	0.3090	0.5879	0.8990	0.9511	1.0	0.5911
0.1	0	0.2939	0.5590	0.7694	0.9045	0.9511	0.9045
0.2	0	0.2500	0.4755	0.6545	0.7694	0.8090	0.7694
0.3	0	0.1816	0.3455	0.4755	0.5590	0.5878	0.5590
0.4	0	0.0955	0.1816	0.2500	0.2939	0.3090	0.2939
0.5	0	0	0	0	0	0	0
0.6	0	−0.0955	−0.1815	−0.2500	−0.2939	−0.3090	−0.2939
0.7	0	−0.1816	−0.3455	−0.4755	−0.5590	−0.4878	−0.5590
⋮	⋮	⋮	⋮	⋮	⋮	⋮	⋮

$t \geq 0$. We can either calculate the values by hand or write a simple computer program. With the MATLAB® code in Figure 3.8, the result shown in Table 3.5 is obtained for $\Delta t = \Delta x = 0.1$. The finite difference solution agrees with the exact, analytic solution in Equation 3.25 to six decimal places. The accuracy of the finite difference solution can be increased by choosing a smaller spatial increment Δx and a smaller time increment Δt.

3.4 Finite Differencing of Elliptic PDEs

A typical elliptic PDE is Poisson's equation, which in two dimensions is given by

$$\nabla^2 \Phi = \frac{\partial^2 \Phi}{\partial x^2} + \frac{\partial^2 \Phi}{\partial y^2} = g(x, y) \tag{3.28}$$

We can use the central difference approximation for the partial derivatives, of which the simplest forms are

$$\frac{\partial^2 \Phi}{\partial x^2} = \frac{\Phi(i+1, j) - 2\Phi(i, j) + \Phi(i-1, j)}{(\Delta x)^2} + O(\Delta x)^2 \tag{3.29a}$$

$$\frac{\partial^2 \Phi}{\partial y^2} = \frac{\Phi(i, j+1) - 2\Phi(i, j) + \Phi(i, j-1)}{(\Delta y)^2} + O(\Delta y)^2 \tag{3.29b}$$

where $x = i\Delta x$, $y = j\Delta y$, and $i,j = 0, 1, 2, ...$ If we assume that $\Delta x = \Delta y = h$ to simplify calculations, substituting Equation 3.29 into Equation 3.28 gives

$$[\Phi(i+1, j) + \Phi(i-1, j) + \Phi(i, j+1) + \Phi(i, j-1)] - 4\Phi(i, j) = h^2 g(i, j)$$

or

$$\Phi(i, j) = \frac{1}{4}[\Phi(i+1, j) + \Phi(i-1, j) + \Phi(i, j+1) + \Phi(i, j-1) - h^2 g(i, j)] \quad (3.30)$$

This is valid at every point (i, j) in the mesh for Poisson's equation. The spatial increment h is called the *mesh size*. A special case of Equation 3.28 is when the source term vanishes, that is, $g(x, y) = 0$. This leads to Laplace's equation. Thus, for Laplace's equation, Equation 3.30 becomes

$$\Phi(i, j) = \frac{1}{4}[\Phi(i+1, j) + \Phi(i-1, j) + \Phi(i, j+1) + \Phi(i, j-1)] \quad (3.31)$$

It is worth noting that Equation 3.31 states that the value of Φ for each point is the average of those at the four surrounding points. The five-point computational molecule for the difference scheme in Equation 3.31 is illustrated in Figure 3.9a, where values of the coefficients are shown. This is a convenient way of displaying finite different algorithms for elliptic PDEs. The molecule in Figure 3.9a is the second-order approximation of Laplace's equation. This

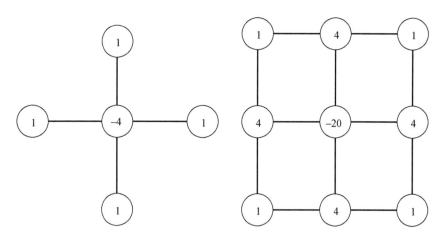

FIGURE 3.9
Computational molecule for Laplace's equation based on (a) second-order approximation, (b) fourth-order approximation.

is obviously not the only way to approximate Laplace's equation, but it is the most popular choice. An alternative fourth-order difference is

$$-20\Phi(i,j)+4[\Phi(i+1,j)+\Phi(i-1,j)+\Phi(i,j+1)+\Phi(i,j-1)]$$
$$+\Phi(i+1,j-1)+\Phi(i-1,j-1)+\Phi(i-1,j+1) \tag{3.32}$$
$$+\Phi(i+1,j+1)=0$$

The corresponding computational molecule is shown in Figure 3.9b.

The application of the finite difference method (FDM) to elliptic PDEs often leads to a system of algebraic equations, and their solution is a major problem in itself. Two commonly used methods of solving the system of equations are band matrix and iterative methods.

3.4.1 Band Matrix Method

From Equations 3.30 to 3.32, we note that only the nearest neighbor nodes affect the value of Φ at each node. Hence, application of any of Equations 3.30 to 3.32 to all the free nodes in the solution region results in a set of simultaneous equations of the form

$$[A][X]=[B] \tag{3.33}$$

where [A] is a sparse matrix (it has many zero elements), [X] is a column matrix consisting of the unknown values of Φ at the free nodes, and [B] is a column matrix containing the known values of Φ at the fixed nodes. Matrix [A] is also banded in that its nonzero terms appear clustered near the main diagonal. Matrix [X], containing the unknown elements, can be obtained from

$$[X]=[A]^{-1}[B] \tag{3.34}$$

In MATLAB, Equation 3.34 can be written as [X] = inv([A]) [B].

3.4.2 Iterative Method

The iterative methods are generally used to solve a large system of simultaneous equations. An iterative method for solving equations is one in which a first approximation is used to calculate a second approximation, which in turn is used to calculate the third approximation, and so on. The three common iterative methods, Jacobi, Gauss–Seidel, and successive over-relaxation (SOR), are mentioned in [2]. We will discuss only SOR here.

To apply the method of SOR to Equation 3.30, for example, we first define the *residual* $R(i,j)$ at node (i,j) as the amount by which the value of $\Phi(i,j)$ does not satisfy Equation 3.30, that is

$$R(i,j) = \Phi(i+1,j) + \Phi(i-1,j) + \Phi(i,j+1)$$

$$+ \Phi(i,j-1) - 4\Phi(i,j) - h^2 g(i,j) \tag{3.35}$$

The value of the residual at the k-th iteration, denoted by $R^k(i,j)$, may be regarded as a correction that must be added to $\Phi(i,j)$ to make it nearer to the correct value. As convergence to the correct value is approached, $R^k(i,j)$ tends to zero. Hence, to improve the rate of convergence, we multiply the residual by a number ω and add that $\Phi(i,j)$ at the k-th iteration to get $\Phi(i,j)$ at $(k+1)$-th iteration. Thus,

$$\Phi^{k+1}(i,j) = \Phi^k(i,j) + \frac{\omega}{4} R^k(i,j)$$

or

$$\Phi^{k+1}(i,j) = \Phi^k(i,j) + \frac{\omega}{4}[\Phi^k(i+1,j) + \Phi^k(i-1,j) + \Phi^k(i,j-1)$$

$$+ \Phi^k(i,j+1) - h^2 g(i,j)] \tag{3.36}$$

The parameter ω is called the *relaxation factor*, and the technique is known as the method of *successive over-relaxation* (SOR). The value of ω lies between 1 and 2. (When $\omega = 1$, the method is simply called *successive relaxation*.) Its optimum value ω_{opt} must be found by trial and error. In order to start solving Equation 3.36, an initial guess $\Phi^0(i,j)$ is made at every free node. Typically, we may choose $\Phi^0(i,j) = 0$ or the average of Φ at the fixed nodes.

Example 3.3
Solve Laplace's equation

$$\nabla^2 V = 0, \quad 0 \le x, y \le 1$$

with $V(x,1) = 45x(1-x), \quad V(x,0) = 0 = V(0,y) = V(1,y)$

Solution
Let $h = 1/3$ so that the solution region is in Figure 3.10. Although the value of h is high, we choose it to allow easy hand calculation. Applying

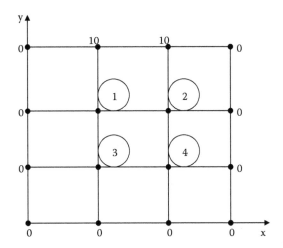

FIGURE 3.10
Finite difference grid for the problem in Example 3.3.

Equation 3.31 to each of the four points leads to

$$4V_1 - V_2 - V_3 - 0 = 10$$

$$-V_1 + 4V_2 - 0 - V_4 = 10$$

$$-V_1 - 0 + 4V_3 - V_4 = 0$$

$$-0 - V_2 - V_3 + 4V_4 = 0$$

This can be written as

$$\begin{bmatrix} 4 & -1 & -1 & 0 \\ -1 & 4 & 0 & -1 \\ -1 & 0 & 4 & -1 \\ 0 & -1 & -1 & 4 \end{bmatrix} \begin{bmatrix} V_1 \\ V_2 \\ V_3 \\ V_4 \end{bmatrix} = \begin{bmatrix} 10 \\ 10 \\ 0 \\ 0 \end{bmatrix}$$

or

$$[A][V] = [B]$$

where [A] is the banded matrix, [V] is the column matrix containing the unknown potentials at the free nodes, and [B] is the column matrix of potentials at the fixed nodes. Solving equation using MATLAB

$$[V] = \text{inv}([A]) \, [B]$$

gives

$$V_1 = 3.75, \quad V_2 = 3.75, \quad V_3 = 1.25, \quad V_4 = 1.25$$

Example 3.4

Solve Poisson's equation

$$\nabla^2 V = -\frac{\rho_s}{\varepsilon}, \quad 0 \le x, y \le 1$$

$\rho_s = x(y-1) \text{ nC/m}^3$ and $\varepsilon_r = 1.0$. The solution region in shown in Figure 3.11.

Solution

This problem has an exact analytic solution and is deliberately chosen so that we can verify the numerical results with exact ones, and we can also see how a problem with complicated analytic solutions is easily solved using FDM.

For the exact solution, we use the superposition theorem and let

$$V = V_L + V_P$$

where V_L is the solution to Laplace's equation $\nabla^2 V_L = 0$ with the inhomogeneous boundary conditions shown in Figure 3.11 and V_P is the solution of

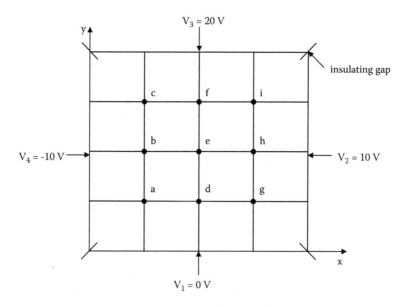

FIGURE 3.11
Solution region for the problem in Example 3.4.

Poisson's equation $\nabla^2 V_P = g = -\rho_s/\varepsilon$ subject to the homogeneous boundary conditions.

$$V_L = V_I + V_{II} + V_{III} + V_{IV}$$

where

$$V_I = \frac{4V_1}{\pi} \sum_{n=\text{odd}}^{\infty} \frac{\sin(n\pi x/a)\sinh[n\pi(b-y)/a]}{n\sinh(n\pi b/a)}$$

$$V_{II} = \frac{4V_2}{\pi} \sum_{n=\text{odd}}^{\infty} \frac{\sin(n\pi x/b)\sinh(n\pi y/b)}{n\sinh(n\pi a/b)}$$

$$V_{III} = \frac{4V_3}{\pi} \sum_{n=\text{odd}}^{\infty} \frac{\sin(n\pi x/a)\sinh(n\pi y/a)}{n\sinh(n\pi b/a)}$$

$$V_{IV} = \frac{4V_4}{\pi} \sum_{n=\text{odd}}^{\infty} \frac{\sin[n\pi(x-a)/b]\sinh(n\pi y/b)}{n\sinh(n\pi a/b)}$$

V_P can be obtained by series expansion:

$$V_P = \sum_{m=1}^{\infty}\sum_{n=1}^{\infty} A_{mn} \sin\frac{m\pi x}{a}\sin\frac{n\pi y}{b}$$

where

$$A_{mn} = \int_0^a \int_0^b g(x,y)\sin\frac{m\pi x}{a}\sin\frac{n\pi y}{b}\,dxdy$$

$$= \frac{[1.0 - \frac{1}{b}[1-(-1)^n]]}{[(m\pi/a)^2 + (n\pi/b)^2]} \cdot \frac{(-1)^{m+n}144ab}{mn\pi}$$

$$a = b = 1 \quad \text{and} \quad g(x,y) = -x(y-1)\bullet 10^{-9}/\varepsilon_o$$

For the finite difference solution, it can be shown that in a rectangular region, the optimum over-relaxation factor is given by the smaller root of the quadratic equation

$$t^2\omega^2 - 16\omega + 16 = 0$$

where $t = \cos(\pi N_x) + \cos(\pi N_y)$ and N_x and N_y are the number of intervals along the x- and y-axes, respectively. Hence,

$$\omega = \frac{8 - \sqrt{64 - 16t^2}}{t^2}$$

We try three cases of $N_x = N_y = 4$, 12, and 20 so that $\Delta x = \Delta y = h = \frac{1}{4}$, $\frac{1}{2}$, and respectively. Also, we set

$$g(x,y) = -\frac{\rho_s}{\varepsilon} = -\frac{x(y-1)10^{-9}}{10^{-9}/36\pi} = -36\pi x(y-1)$$

Figure 3.12 presents the MATLAB code for the solution of the problem. The potentials at the free nodes for different cases of h are shown in Table 3.6.

```
%%%%%%%%%%%%%%%%%%%%%%%%%%%%%%%%%%%%%%%%
%   FINITE DIFFERENCE SOLUTION OF POISSON'S EQUATION:
%            Vxx + Vyy = G
%   USING THE METHOD OF SUCCESSIVE OVER-RELAXATION
%
%   NX   : NO. OF INTERVALS ALONG X-AXIS
%   NY   : NO. OF INTERVALS ALONG Y-AXIS
%   A X B : DIMENSION OF THE SOLUTION REGION
%   V(I,J) : POTENTIAL AT GRID POINT (X,Y) = H*(I,J)
%          WHERE I = 0,1,...,NX, J = 0,1,....,NY
%   H    : MESH SIZE
% ************************************         *** ****************

    % SPECIFY BOUNDARY VALUES AND NECESSARY PARAMETERS
    A=1;B=1;
    V1=0;V2=10;V3=20;V4=-10;
    NX= 20; %4 12 20
    NY= NX;
    H = A/NX;
    % SET INITIAL GUESS EQUAL TO ZEROS OR TO AVERAGE
     %  OF FIXED VALUES

    for I=1:NX-1
      for J=1:NY-1
        V(I+1,J+1)=(V1 + V2 + V3 + V4)/4.0;
      end
    end
    % SET POTENTIALS AT FIXED NODES
    for I = 1:NX-1
      V(I+1,1)=V1;
      V(I+1,NY+1)=V3;
    end
    for J=1:NY-1
      V(1,J+1)=V4;
      V(NX+1,J+1)=V2;
    end
    V(1,1)=(V1 + V4)/2.0;
    V(NX+1,1)=(V1 + V2)/2.0;
    V(1,NY+1)=(V3 + V4)/2.0;
    V(NX+1,NY+1)=(V2 + V3)/2.0;
    % FIND THE OPTIMUM OVER-RELAXATION FACTOR
```

FIGURE 3.12
MATLAB code for Example 3.4.

```
T = cos(pi/NX) + cos(pi/NY);
W = ( 8 - sqrt(64 - 16*T^2))/(T^2);
disp(['SOR Factor Omega = ',num2str(W)])
W4 = W/4;
% ITERATION BEGINS
NCOUNT = 0;

loop = 1;
while loop == 1;
   RMIN = 0;
   for I =1:NX-1
     X = H*I;
     for J = 1:NY-1
        Y = H*J;
        G = -36.0*pi*X*(Y - 1.0);
        R = W4*( V(I+2,J+1) + V(I,J+1) + V(I+1,J+2) + V(I+1,J)-4.0*V(I+1,J+1) -
G*H*H  );
           RMIN = RMIN + abs(R);
           V(I+1,J+1) =  V(I+1,J+1) + R;
        end
     end
     RMIN = RMIN/(NX*NY);
     if(RMIN>=0.0001)
        NCOUNT = NCOUNT + 1;
        if(NCOUNT>100)
          loop = 0;
          disp('SOLUTION DOES NOT CONVERGE IN 100 ITERATIONS')
        end
     else
        % Then RMIN is less than .0001 and then solution has converged
        loop = 0;
        disp(['Solution Converges in ',num2str(NCOUNT),' iterations'])
        disp(['h = ', num2str(H)])
     end
   end

Vnum = V;

% Grab original points a through i
     abc = zeros(1,9);
     a_tic = 1;
     vec = [0:H:1];
     for ii = .25:.25:.75
       for jj = .25:.25:.75
          xind = find(vec==ii);
          yind = find(vec==jj);
```

FIGURE 3.12
(Continued).

```
      % disp([xind,yind])
      abc(a_tic) = Vnum(xind,yind);
      a_tic = a_tic + 1;
    end
  end


%     OUTPUT THE FINITIE DIFFERENCE APPROX. RESULTS

%     --------------------------------------------------------
%     CALCULATE THE EXACT SOLUTION
%
%     POISSON'S EQUATION WITH HOMOGENEOUS BOUNDARY CONDITIONS

%     SOLVED BY SERIES EXPANSION
%
  for I =1:NX-1
    X = H*I;
    for J = 1:NY-1
      Y = H*J;
      SUM = 0;
      for M = 1:10  % TAKE ONLY 10 TERMS OF THE SERIES
        FM = M;
        for N = 1:10
          FN = N;
          FACTOR1 = (FM*pi/A)^2 + (FN*pi/B)^2;
          FACTOR2 = ( (-1)^(M+N) )*144*A *B/(pi*F M*F N);
          FACTOR3 = 1 - (1 - (-1)^N)/B;
          FACTOR = FACTOR2*F ACTOR3/FACTOR1;
          SUM = SUM + FACTOR*sin(FM*pi*X /A)*sin(FN*pi*Y /B);
        end
      end
      VH = SUM;

%     LAPLACE'S EQUATION WITH INHOMOGENEOUS BOUNDARY
CONDITIONS
%     SOLVED USING THE METHOD OF SEPARATION OF VARIABLES

      C1=4*V 1/pi;
      C2=4*V 2/pi;
      C3=4*V 3/pi;
      C4=4*V 4/pi;
      SUM=0;
      for K =1:10  % TAKE ONLY 10 TERMS OF THE SERIES
        N=2*K -1;
        AN=N;
```

FIGURE 3.12
(Continued).

```
            A1=sin(AN*pi*X/B);
            A2=sinh(AN*pi*(A-Y)/B);
            A3=AN*sinh(AN*pi*A/B);
            TERM1=C1*A1*A2/A3;
            B1=sinh(AN*pi*X/A);
            B2=sin(AN*pi*Y/A);
            B3=AN*sinh(AN*pi*B/A);
            TERM2=C2*B1*B2/B3;
            D1=sin(AN*pi*X/B);
            D2=sinh(AN*pi*Y/B);
            D3=AN*sinh(AN*pi*A/B);
            TERM3=C3*D1*D2/D3;
            E1=sinh(AN*pi*(B-X)/A);
            E2=sin(AN*pi*Y/A);
            E3=AN*sinh(AN*pi*B/A);
            TERM4=C4*E1*E2/E3;
            TERM = TERM1 + TERM2 + TERM3 + TERM4;
            SUM=SUM + TERM;
        end
        VI = SUM;
        Vexact(I+1,J+1) = VH + VI;
      end
    end

%Grab original points a through i
    abc2 = zeros(1,9);
    a_tic = 1;
    vec = [0:H:1];
    for ii = .25:.25:.75
      for jj = .25:.25:.75
        xind = find(vec==ii);
        yind = find(vec==jj);
        %disp([xind,yind])
        abc2(a_tic) = Vexact(xind,yind);
        a_tic = a_tic + 1;
      end
    end

figure(1),
    imagesc(flipud(Vnum')),
    colorbar
    ylabel('y'),    xlabel('x')
    title('Example 3.4: Poisson PDE')


format short g
    disp('   numerical    exact')
    disp([abc' abc2'])
```

FIGURE 3.12
(Continued).

TABLE 3.6

Successive Over-Relaxation Solution of Example 3.4

Node	$h = 1/4$ $\omega_{opt} = 1.1718$ Iterations	$h = 1/2$ $\omega_{opt} = 1.729$ 26 Iterations	$h = 1/20$ $\omega_{opt} = 1.729$ 43 Iterations	Exact Solution
a	−3.247	−3.409	−3.424	−3.429
b	−1.703	−1.982	−2.012	−2.029
c	4.308	4.279	4.277	−4.277
d	−0.0393	−0.0961	−0.1087	−0.1182
e	3.012	2.928	2.921	2.913
f	9.368	9.556	9.578	9.593
g	3.068	2.921	2.909	2.902
h	6.111	6.072	6.069	6.065
i	11.04	11.12	11.23	11.13

Note that as the mesh size h reduces, the solution becomes more accurate, but it takes more iterations for the same tolerance.

3.5   Accuracy and Stability of Finite Difference Solutions

The question of accuracy and stability of numerical methods is extremely important if our solution is to be reliable and useful. This is especially important for parabolic and hyperbolic PDEs. Accuracy has to do with the closeness of the approximate solution to the exact solutions (assuming that they exist). Stability is the requirement that the scheme does not increase the magnitude of the solution with increase in time.

There are three sources of errors that are nearly unavoidable in numerical solution of physical problems:

- Modeling errors
- Truncation (or discretization) errors
- Round-off errors

Each of these error types will affect accuracy and thereby degrade the solution.

Modeling errors are due to several assumptions made in arriving at the mathematical model. For example, a nonlinear system may be represented by a linear PDE. Truncation errors arise from the fact that in numerical analysis, we can deal only with a finite number of terms from processes that are usually described by infinite series. For example, in deriving finite difference schemes, some higher-order terms in the Taylor series expansion were neglected, thereby introducing truncation error. Truncation errors may be reduced by using finer meshes, that is, by reducing the mesh size h and time increment Δt. Alternatively, truncation errors may be reduced by using a large number of terms in the series expansion of derivatives, that is, by using higher-order approximation. However, care must be exercised in applying higher-order approximation. Instability may result if we apply a difference equation of an order higher than the PDE being examined. These higher-order difference equations may introduce spurious solutions.

Round-off errors reflect the fact that computations can be done only with a finite precision on a computer. This unavoidable source of errors is due to the limited size of registers in the arithmetic unit of the computer. Round-off errors can be minimized by the use of double-precision arithmetic. The only way to avoid round-off errors completely is to code all operations using integer arithmetic. This is hardly possible in most practical situations.

Although it has been noted that reducing the mesh size h will increase accuracy, it is not possible to indefinitely reduce h. Decreasing the truncation error by using a finer mesh may result in increasing the round-off error owing to the increased number of arithmetic operations. A point is reached at which the minimum total error occurs for any particular algorithm using any given word length. This is illustrated in Figure 3.2. Concerns about accuracy lead us to question whether the finite difference solution can grow unbounded, a property termed the *instability* of the difference scheme. A numerical algorithm is said to be stable if a small error at any stage produces a smaller cumulative error. It is unstable otherwise. The consequence of instability (producing unbounded solution) is disastrous. To determine whether a finite difference scheme is stable, we define an error, ε^n, which occurs at time step n, assuming that there is one independent variable. We define the amplification of this error at time step $n + 1$ as

$$\varepsilon^{n+1} = g\varepsilon^n \tag{3.37}$$

where g is known as the *amplification factor*. In a more complex situation, we have two or more independent variables, and Equation 3.37 becomes

$$[\varepsilon]^{n+1} = [G][\varepsilon]^n \tag{3.38}$$

where $[G]$ is the amplification matrix. For the stability of the difference scheme, it is required that Equation 3.37 satisfy

$$|\varepsilon^{n+1}| \le |\varepsilon^n|$$

or

$$|g| \le 1 \tag{3.39a}$$

For the case in Equation 3.38,

$$\|G\| \le 1 \tag{3.39b}$$

One useful and simple method of finding a stability criterion for a difference scheme is to construct a Fourier analysis of the difference equation and thereby derive the amplification factor. We illustrate this technique, known as *von Neumann's method,* by considering the explicit scheme of Equation 3.13:

$$\Phi_i^{n+1} = (1 - 2r)\Phi_i^n + r\left(\Phi_{i+1}^n + \Phi_{i-1}^n\right) \tag{3.40}$$

where $r = \Delta t/k(\Delta x)^2$. We have changed our usual notation so that we can use $j = \sqrt{-1}$ in the Fourier series. Let the solution be

$$\Phi_i^n = \sum A^n(t)e^{jkix}, \quad 0 \le x \le 1 \tag{3.41a}$$

where k is the wave number. Because the differential equation (Equation 3.10) approximated by Equation 3.13 is linear, we need consider only a Fourier mode, that is,

$$\Phi_i^n = A^n(t)e^{jkix} \tag{3.41b}$$

Substituting Equation 3.41b into Equation 3.40 gives

$$A^{n+1}e^{jkix} = (1 - 2r)A^n e^{jkix} + r(e^{jkx} + e^{-jkx})A^n e^{jkix}$$

or

$$A^{n+1} = A^n[1 - 2r - 2r \cos kx]$$

Hence, the amplification factor is obtained as

$$g = \frac{A^{n+1}}{A^n} = 1 - 2r + 2r \cos kx$$

$$= 1 - 4r \sin^2 \frac{kx}{2} \tag{3.43}$$

In order to satisfy Equation 3.39a

$$\left| 1 - 4r \sin^2 \frac{kx}{2} \right| \le 1$$

Because the condition must hold for every wave number k, we take the minimum value of the sine function so that

$$1 - 4r \ge -1 \quad \text{and} \quad r \ge 0$$

or

$$r \ge \frac{1}{2} \quad \text{and} \quad r \ge 0$$

Of course, $r = 0$ implies $\Delta t = 0$, which is impractical. Thus

$$0 < r \le \frac{1}{2}$$

Example 3.5

For the finite difference scheme of Equation 3.20, use the von Neumann approach to determine the stability condition.

Solution

We assume a trial solution of the form

$$\Phi_i^n = A^n(t)e^{jkix}$$

Substituting this into Equation 3.20 results in

$$A^{n+1}e^{jkix} = 2(1-r)A^ne^{kjix} + r(e^{jkx} + e^{-jkx})A^ne^{jkix} - A^{n-1}e^{jkix}$$

or

$$A^{n+1} = A^n[2(1-r)+2r\cos kx] - A^{n-1} \tag{3.45}$$

In terms of $g = A^{n+1}/A^n$, Equation 3.45 becomes

$$g^2 - 2pg + 1 = 0 \tag{3.46}$$

where $p = 1 - 2r\sin^2\frac{kx}{2}$. The quadratic equation (Equation 3.46) has solutions

$$g_1 = p + [p^2 - 1]^{1/2}, \quad g_2 = p - [p^2 - 1]^{1/2}$$

For $|g_i|$, where $i = 1, 2$, p must lie between 1 and -1, that is, $-1 \leq p \leq 1$ or

$$-1 \leq 1 - 2r\sin^2\frac{kx}{2} \leq 1$$

which implies that $r \leq 1$ or $u\Delta t \leq \Delta x$ for stability. This can be extended to show that the stability condition for two-dimensional wave equation is $u\Delta t/h < \frac{1}{\sqrt{2}}$, where $h = \Delta x = \Delta y$.

3.6 Maxwell's Equations

The most common approach to the finite difference solution of Maxwell's equations is the finite difference time domain (FDTD) method. This was first introduced by Yee in 1966 and later developed by Taflove and others [5–7]. In this method, the Maxwell's curl equations are rewritten using finite difference notation. The principal idea in this method is to employ time- and space-centered approximation for the derivatives occurring in the Maxwell's equations with a second-order accuracy. In an isotropic medium, Maxwell's equations can be written as

$$\nabla \times E = -\mu \frac{\partial H}{\partial t} \tag{3.47a}$$

$$\nabla \times H = \sigma E + \varepsilon \frac{\partial E}{\partial t} \tag{3.47b}$$

Written component-wise, these correspond to six scalar equations. Following Yee's notation, we define a grid point in the solution region as

$$(i, j, k) \equiv (i\Delta x, j\Delta y, k\Delta z) \tag{3.48}$$

and a function F of space and time at n-th time instant as

$$F^n(i, j, k) = F(i\delta, j\delta, k\delta, n\Delta t) \tag{3.49}$$

where $\delta = \Delta x = \Delta y = \Delta z$ is the space increment, Δt is the time increment, and i, j, k, and n are integers. Using central difference approximations for space and time derivatives that are second-order accurate,

$$\frac{\partial F^n(i, j, k)}{\partial x} = \frac{F^n(i+1/2, j, k) - F^n(i-1/2, j, k)}{\delta} + O(\delta^2) \tag{3.50}$$

$$\frac{\partial F^n(i, j, k)}{\partial t} = \frac{F^{n+1/2}(i, j, k) - F^{n-1/2}(i, j, k)}{\Delta t} + O(\Delta t^2) \tag{3.51}$$

In applying Equation 3.50 to all the space derivatives in Equations 3.47, Yee positions the components of **E** and **H** about a unit cell of lattice as shown in Figure 3.13. To incorporate Equation 3.51, the components of **E** and **H** are evaluated at alternate half-time steps. Thus, we obtain the explicit finite difference approximations of Equations 3.47a and 3.47b as

$$H_x^{n+1/2}(i, j+1/2, k+1/2) = H_x^{n-1/2}(i, j+1/2, k+1/2) + \frac{\Delta t}{\mu(i, j+1/2, k+1/2)\delta}$$
$$\times \left[\begin{array}{l} E_y^n(i, j+1/2, k+1) - E_y^n(i, j+1/2, k) \\ + E_z^n(i, j, k+1/2) - E_z^n(i, j+1, k+1/2) \end{array} \right] \tag{3.52a}$$

$$H_y^{n+1/2}(i+1/2, j, k+1/2) = H_y^{n-1/2}(i+1/2, j, k+1/2) + \frac{\Delta t}{\mu(i+1/2, j, k+1/2)\delta}$$
$$\times \left[\begin{array}{l} E_z^n(i+1, j, k+1/2) - E_z^n(i, j, k+1/2) \\ + E_x^n(i+1/2, j, k) - E_x^n(i+1/2, j, k+1) \end{array} \right] \tag{3.52b}$$

$$H_z^{n+1/2}(i+1/2, j+1/2, k) = H_z^{n-1/2}(i+1/2, j+1/2, k) + \frac{\Delta t}{\mu(i+1/2, j+1/2, k)\delta}$$
$$\times \left[\begin{array}{l} E_x^n(i+1/2, j+1, k) - E_x^n(i+1/2, j, k) \\ + E_y^n(i, j+1/2, k) - E_y^n(i+1, j+1/2, k) \end{array} \right] \tag{3.52c}$$

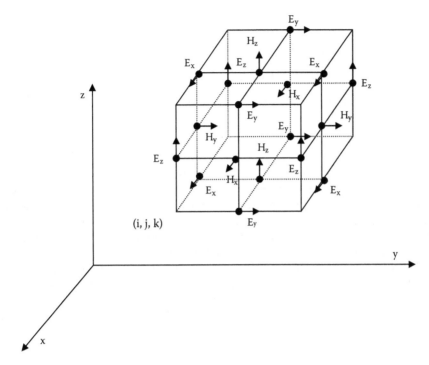

FIGURE 3.13
Positions of the field components in the Yee cell.

$$E_x^{n+1}(i+1/2,j,k)=\left(1-\frac{\sigma(i+1/2,j,k)\Delta t}{\varepsilon(i+1/2,j,k)}\right)E_x^n(i+1/2,j,k)$$

$$+\frac{\Delta t}{\varepsilon(i+1/2,j,k)\delta}\left[\begin{array}{l}H_z^{n+1/2}(i+1/2,j+1/2,k)-H_z^{n+1/2}(i+1/2,j-1/2,k)\\+H_y^{n+1/2}(i+1/2,j,k-1/2)-H_y^{n+1/2}(i+1/2,j,k+1/2)\end{array}\right]$$

(3.52d)

$$E_y^{n+1}(i,j+1/2,k)=\left(1-\frac{\sigma(i,j+1/2,k)\Delta t}{\varepsilon(i,j+1/2,k)}\right)E_y^n(i,j+1/2,k)$$

$$+\frac{\Delta t}{\varepsilon(i,j+1/2,k)\delta}\left[\begin{array}{l}H_x^{n+1/2}(i,j+1/2,k+1/2)-H_x^{n+1/2}(i,j+1/2,k-1/2)\\+H_z^{n+1/2}(i-1/2,j+1/2,k)-H_z^{n+1/2}(i+1/2,j+1/2,k)\end{array}\right]$$

(3.52e)

$$E_z^{n+1}(i,j,k+1/2) = \left(1 - \frac{\sigma(i,j,k+1/2)\Delta t}{\varepsilon(i,j,k+1/2)}\right)E_z^n(i,j,k+1/2)$$

$$+ \frac{\Delta t}{\varepsilon(i,j,k+1/2)\delta}\left[\begin{array}{l} H_y^{n+1/2}(i+1/2,j,k+1/2) - H_y^{n+1/2}(i-1/2,j,k+1/2) \\ + H_x^{n+1/2}(i,j-1/2,k+1/2) - H_x^{n+1/2}(i,j+1/2,k+1/2) \end{array}\right]$$

(3.52f)

Note from Equations 3.52a–f that the components of **E** and **H** are interlaced within the unit cell and are evaluated at alternate half-time steps. All the field components are present in a quarter of a unit cell. In translating the system of Equations 3.52a–f into computer code, one must make sure that, within the same time loop, one type of field component is calculated first and the results obtained are used in calculating another type.

To ensure accuracy of the computed results, spatial increment δ must be small compared to the wavelength (usually $\leq \lambda/10$) or minimum dimension of the conducting body. This amounts to having 10 or more cells per wavelength. To ensure the stability of the FDTD scheme, the time increment must satisfy

$$u_{max}\Delta t \leq \left[\frac{1}{\Delta x^2} + \frac{1}{\Delta y^2} + \frac{1}{\Delta z^2}\right]^{-1/2}$$

(3.53)

where u_{max} is the maximum phase velocity within the model. Equation 3.53 is called *Courant stability condition*. Because we are using a cubic cell with $\Delta x = \Delta y = \Delta z = \delta$, Equation 3.53 becomes

$$\frac{u_{max}\Delta t}{\delta} \leq \frac{1}{\sqrt{n}}$$

(3.54)

where n is the number of space dimensions. For practical reasons, it is good to choose the ratio of time increment to spatial increment as large as possible while satisfying Equation 3.54.

A basic difficulty encountered in applying the FDTD method to problems involving open or unbounded geometries is that in the domain the field is infinite. Because no computer can store unlimited data, we should devise a scheme to confine our solution region to a finite extent. In other words, artificial boundaries must be simulated that create a numerical illusion. The conditions under which this can happen are called *radiation conditions, absorbing boundary conditions,* or *lattice truncation conditions.*

The FDTD method has several advantages. It is conceptually simple. It can treat even inhomogeneous conducting or dielectric structures of complex geometries, because the medium parameters can be assigned to each lattice point. It also permits computation of the solution in the frequency domain from a single time domain simulation via Fourier transform techniques. A disadvantage of the method is that it necessitates modeling of an object as well as its surroundings.

Thus, the program execution time may be excessive. Its accuracy is at least one order of magnitude worse than that of the moment method. In spite of these limitations, the FDTD method has been successfully applied to solve scattering and other problems including aperture penetration, antenna/ radiation problems, microwave circuits, eigenvalue problems, electromagnetic (EM) absorption in human tissues, etc.

3.7 Summary

The finite difference method is a very versatile means of solving PDEs commonly encountered in electromagnetics. Finite difference has been applied to parabolic, hyperbolic, and elliptic equations. The problems of stability and convergence of the finite difference solution have been discussed. For solving Maxwell's equation, the components of **E** and **H** are evaluated at alternate half-time steps.

References

1. M. N. O. Sadiku and S. R. Nelatury, Computational electromagnetics, in R. Dorf (Ed.), *Electrical Engineering Handbook*, 3rd ed., Boca Raton, FL: CRC Press, 2006, pp. 23.1–23.26.
2. M. N. O. Sadiku, *Numerical Techniques in Electromagnetics*, 3rd ed., Boca Raton, FL: CRC Press, 2009.
3. R. D. Richtmyer and K. W. Morton, *Difference Methods for Initial-Value Problems*, 2nd ed., New York: Interscience Publ., 1976, pp. 185–193.
4. D. Potter, *Computational Physics*, London: John Wiley, 1973, pp. 40–79.
5. K. S. Yee, Numerical solution of initial boundary value problems involving Maxwell's equations in isotropic media, *IEEE Trans. Antenn. Propag.*, vol. 14, 302–307, 1966.
6. K. S. Kunz and R. J. Luebbers, *The Finite Difference Time Domain Method for Electromagnetics*, Boca Raton, FL: CRC Press, 1993.
7. A. Taflove, *Computational Electrodynamics: The Finite-Difference Time: Domain Method*, Boston: Artech House, 1995.

4

Fixed Random Walk

For want of a nail the shoe was lost; for want of a shoe the horse was lost; for want of a horse the ride was lost.

—Proverb

4.1 Introduction

Armed with random numbers and statistical methods of Monte Carlo simulations, we can now apply Monte Carlo techniques to physical problems. We begin with potential problems, which are among the most interesting areas where Monte Carlo methods (MCMs) have found applications.

As mentioned in Chapter 1, the connection between potential theory and Brownian motion (or random walk) was first shown in 1944 by Kakutani [1]. Subsequent investigators [2–4] have extended the work of Kakutani. The first systematic application of MCMs to electromagnetic problems was that of Royer [5]. A simple introduction to the subject is given in [6].

An underlying concept of the probabilistic or Monte Carlo solution of differential equations is the *random walk*. Different types of random walk lead to different MCMs. The most popular types include the *fixed random walk, nonreversal random walk, self-avoiding random walk, floating random walk,* and *fixed-radius floating random walk*. The nonreversal and self-avoiding random walks are different versions of the fixed random walk [7, 8], whereas the floating random walk is a completely different type of MCM. The fixed random walk will be discussed in this chapter, whereas the floating random walk will be treated in Chapter 5.

The application of the fixed random walk to potential problems usually involves three steps:

1. Obtain the transient probabilities from the finite difference equivalent of the partial differential equation describing the problem.
2. Use random numbers along with the transient probabilities to direct several random walks in the solution region and record the potential at the end of each random walk.
3. Find the statistical average of the recorded potentials in step 2.

4.2 Solution of Laplace's Equation

Let us suppose that the fixed random walk is to be applied to solve Laplace's equation

$$\nabla^2 V = 0 \quad \text{in region R} \tag{4.1a}$$

or

$$0 = \frac{\partial^2 V}{\partial x^2} + \frac{\partial^2 V}{\partial y^2} + \frac{\partial^2 V}{\partial z^2} \tag{4.1b}$$

subject to the Dirichlet boundary condition

$$V = V_p \quad \text{on boundary B} \tag{4.2}$$

We first consider the one-dimensional case.

4.2.1 One-Dimensional Case

We divide R into equal subintervals of length Δ. The central finite difference approximations of the first and second derivatives of V with respect to x are [9]

$$\frac{dV}{dx} = V' \simeq \frac{V(x+\Delta) - V(x)}{\Delta} \tag{4.3}$$

$$\frac{d^2 V}{dx^2} \simeq \frac{V'(x+\Delta/2) - V'(x-\Delta/2)}{\Delta}$$

$$= \frac{V(x+\Delta) - 2V(x) + v(x-\Delta)}{\Delta^2} \tag{4.4}$$

Substituting Equation 4.4 into Equation 4.1 gives the finite difference approximation of Laplace's equation, namely,

$$0 = \frac{V(x+\Delta) - 2V(x) + v(x-\Delta)}{\Delta^2}$$

or

$$V(x) = p_{x+} V(x+\Delta) + p_{x-} V(x-\Delta) \tag{4.5}$$

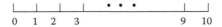

FIGURE 4.1
One-dimensional fixed random walk.

where

$$p_{x+} = p_{x-} = \frac{1}{2} \tag{4.6}$$

p_{x+} and p_{x-} are known as *transient probabilities* or *random walk probabilities*. A probabilistic interpretation may be given to Equation 4.5. A random-walking particle at point x moves to new location $x + \Delta$ with probability 1/2 or to $x - \Delta$ with probability 1/2. This suggests that by generating a random number U uniformly distributed in the interval (0,1), we can instruct the particle to move as follows:

$$x \rightarrow x + \Delta \quad \text{if} \quad 0 < U < 0.5$$

$$x \rightarrow x - \Delta \quad \text{if} \quad 0.5 < U < 1 \tag{4.7}$$

An interesting illustration of random walks in one dimension is a random walk along Madison Avenue [10]. A man walks along a 10-block stretch of Madison Avenue as shown in Figure 4.1. Starting at node k, he walks one block to the right or left with probability 1/2. He wanders until he reaches node 0, which is a bar, or node 10, which is home. When he reaches either the bar or home, he stays there. The problem is, starting at node k, what is the probability $p(k)$ that the man will reach home before reaching the bar? The answers turns out to be $p(k) = k/10$. The Monte Carlo solution to this problem consists in sending out N random-walking particles from point k, directing their movement using Equation 4.7, and counting how many of them reach node 10 before being absorbed (see Problem 4.1).

4.2.2 Two-Dimensional Case

We first divide the solution region R into mesh and replace ∇^2 by its finite difference equivalent. From Equation 4.4,

$$\frac{\partial^2 V}{\partial x^2} \approx \frac{V(x + \Delta x, y) - 2V(x, y) + V(x - \Delta x, y)}{\Delta x^2} \tag{4.8a}$$

$$\frac{\partial^2 V}{\partial y^2} \approx \frac{V(x, y + \Delta y) - 2V(x, y) + V(x, y - \Delta y)}{\Delta y^2} \tag{4.8b}$$

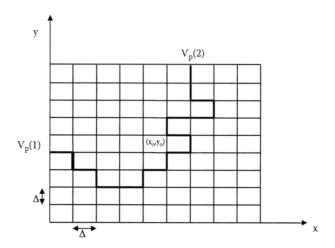

FIGURE 4.2
Two-dimensional fixed random walks.

The finite difference representation of Equation 4.1 in two-dimensional R is obtained by substituting Equation 4.8 into Equation 4.1. The result is

$$0 = [V(x+\Delta,y)+V(x-\Delta,y)+V(x,y+\Delta)+V(x,y-\Delta)-4V(x,y)]/\Delta^2$$

where a square grid of mesh size $\Delta = \Delta x = \Delta y$, such as in Figure 4.2, has been assumed. Solving for $V(x,y)$ yields

$$V(x,y) = p_{x+}V(x+\Delta,y)+p_{x-}V(x-\Delta,y)+p_{y+}V(x,y+\Delta)+p_{y-}V(x,y-\Delta) \quad (4.9)$$

where

$$p_{x+} = p_{x-} = p_{y+} = p_{y-} = \frac{1}{4} \tag{4.10}$$

Equation 4.9 may be given a probabilistic interpretation. If a random-walking particle is instantaneously at the point (x, y), it has transient probabilities p_{x+}, p_{x-}, p_{y+}, and p_{y-} of moving from (x, y) to point $(x + \Delta, y)$, $(x - \Delta, y)$, $(x, y + \Delta)$, and $(x, y - \Delta)$, respectively. A means of determining which way the particle should move is to generate a uniformly distributed random number U, $0 < U < 1$, and instruct the particle to walk as follows:

$$(x,y) \rightarrow (x+\Delta,y) \quad \text{if} \quad 0 < U < 0.25$$

$$(x,y) \rightarrow (x-\Delta,y) \quad \text{if} \quad 0.25 < U < 0.5$$

$$(x,y) \rightarrow (x,y+\Delta) \quad \text{if} \quad 0.5 < U < 0.75$$

$$(x,y) \rightarrow (x,y-\Delta,) \quad \text{if} \quad 0.75 < U < 1 \tag{4.11}$$

If a rectangular grid rather than a square grid is employed, then $p_{x+} = p_{x-} = p_x$ and $p_{y+} = p_{y-} = p_y$, but $p_x \neq p_y$.

To calculate the potential at (x_o, y_o), a random-walking particle is instructed to start at that point. The particle proceeds to wander from node to node in the grid until it reaches the boundary. When it does, the walk is terminated, and the prescribed potential V_p at that boundary point is recorded. Let the value of V_p at the end of the first walk be denoted by V_p (1), as illustrated in Figure 4.2. Then a second particle is released from (x_o, y_o) and allowed to wander until it reaches a boundary point, where the walk is terminated and the corresponding value of V_p is recorded as V_p (2). This procedure is repeated for the third, fourth, ..., and N-th particle released from (x_o, y_o) and the corresponding prescribed potential V_p (3), V_p (4), ..., V_p (N) are noted. According to Kakutani [1], the expected value of V_p (1), V_p (2), ..., V_p (N) is the solution of the Dirichlet problem at (x_o, y_o), that is,

$$V(x_o, y_o) = \frac{1}{N} \sum_{i=1}^{N} V_p(i) \tag{4.12}$$

where N, the total number of walks, is large. The rate of convergence varies as $\sqrt{N}$ so that many random walks are required to ensure accurate results.

An interesting analogy to the two-dimensional random walk is the walking drunk problem [11,12]. We regard the random-walking particle as a "drunk," the squares of the mesh as the "blocks in a city," the nodes as "crossroads," the boundary B as the "city limits," and the terminus on B as the "policeman." Though the drunk is trying to walk home, he is so intoxicated that he wanders randomly throughout the city. The job of the policeman is to seize the drunk in his first appearance at the city limits and ask him to pay a fine V_p. What is the expected fine the drunk will receive? The answer to this question is in Equation 4.12.

On the dielectric boundary, the boundary condition $D_{1n} = D_{2n}$ is imposed. Consider the interface along y = constant plane as shown in Figure 4.3.

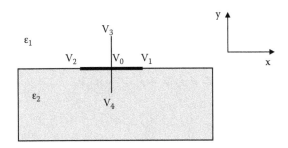

FIGURE 4.3
Interface between media of dielectric permittivities ε_1 and ε_2.

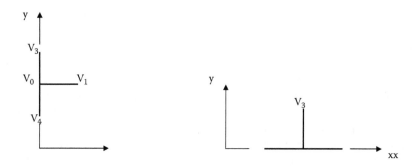

FIGURE 4.4
Satisfying symmetry conditions: (a)

According to [9, 13], the finite difference equivalent of the boundary condition at the interface is

$$V_o = p_{x+}V_1 + p_{x-}V_2 + p_{y+}V_3 + p_{y-}V_4 \tag{4.13}$$

where the transient probabilities are given by

$$p_{x+} = p_{x-} = \frac{1}{4}, \quad p_{y+} = \frac{\varepsilon_1}{2(\varepsilon_1 + \varepsilon_2)}, \quad p_{y-} = \frac{\varepsilon_2}{2(\varepsilon_1 + \varepsilon_2)} \tag{4.14}$$

An interface along an $x = $ constant plane can be treated in a similar manner.
   On a line of symmetry, the condition $\frac{\partial V}{\partial n} = 0$ must be imposed. If the line of symmetry is along the y-axis as in Figure 4.4a, according to [9]

$$V_o = p_{x+}V_1 + p_{y+}V_3 + p_{y-}V_4 \tag{4.15}$$

where

$$p_{x+} = \frac{1}{2}, \quad p_{y+} = p_{y-} = \frac{1}{4} \tag{4.16}$$

The line of symmetry along the x-axis, shown in Figure 4.4b, is treated similarly, that is,

$$V_o = p_{x+}V_1 + p_{x-}V_2 + p_{y+}V_3 \tag{4.17}$$

where

$$p_{x+} = p_{x-} = \frac{1}{4}, \quad p_{y+} = \frac{1}{2} \tag{4.18}$$

In each of the cases in Equations 4.14, 4.16, and 4.18, the uniformly distributed random number U is subdivided in proportion to the transient probabilities and the random walker is instructed to move accordingly.

4.2.3 Three-Dimensional Case

For a three-dimensional problem in which cubical cells are used, the transient probabilities are easily obtained as

$$P_{x+} = P_{x-} = P_{y+} = P_{y-} = P_{z+} = P_{z-} = \frac{1}{6} \qquad (4.19)$$

In this case, the random number U is subdivided according to these probabilities, and the random-walking particles are directed accordingly.

Example 4.1

Give a probabilistic interpretation using the finite difference form of the energy equation

$$u \frac{\partial T}{\partial x} + v \frac{\partial T}{\partial y} = \alpha \left(\frac{\partial^2 T}{\partial x^2} + \frac{\partial^2 T}{\partial y^2} \right)$$

Assume a square grid of size Δ.

Solution

If a backward difference is applied to the left-hand side, whereas a forward difference to the right-hand side, we obtain

$$u \frac{T(x,y) - T(x-\Delta,y)}{\Delta}$$

$$+ v \frac{T(x,y) - T(x,y-\Delta)}{\Delta} = \alpha \frac{T(x+\Delta,y) - 2T(x,y) + T(x-\Delta,y)}{\Delta^2}$$

$$+ \alpha \frac{T(x,y+\Delta) - 2T(x,y) + T(x,y-\Delta)}{\Delta^2} \qquad (4.20)$$

Rearranging terms leads to

$$T(x,y) = p_{x+} T(x+\Delta,y) + p_{x-} T(x-\Delta,y) + p_{y+} T(x,y+\Delta) + p_{y-} T(x,y-\Delta) \qquad (4.21)$$

where

$$p_{x+} = p_{y+} = \frac{1}{\frac{u\Delta}{\alpha} + \frac{v\Delta}{\alpha} + 4} \qquad (4.22a)$$

$$p_{x-} = \frac{\left(1 + \frac{\Delta u}{\alpha}\right)}{\frac{u\Delta}{\alpha} + \frac{v\Delta}{\alpha} + 4} \qquad (4.22b)$$

$$p_{x-} = \frac{\left(1 + \frac{\Delta v}{\alpha}\right)}{\frac{u\Delta}{\alpha} + \frac{v\Delta}{\alpha} + 4} \qquad (4.22c)$$

Equation 4.21 is given a probabilistic interpretation as follows: a walker at point (x, y) has probabilities $p_{x+}, p_{x-}, p_{y+},$ and p_{y-} of moving from (x, y) to point $(x + \Delta, y)$, $(x - \Delta, y)$, $(x, y + \Delta)$, and $(x, y - \Delta)$, respectively. With this interpretation, Equation 4.21 can be used to solve the differential equation with fixed random MCM.

Example 4.2

Consider a conducting trough of infinite length with the square cross section shown in Figure 4.5. The trough wall at $y = 1$ is connected to 100 V, whereas the other walls are grounded as shown. We intend to find the potential within the trough using the fixed random walk MCM.

Solution

The problem is solving Laplace's equation subject to

$$V(0, y) = V(1, y) = V(x, 0) = 0, \quad V(x, 1) = 100 \qquad (4.23)$$

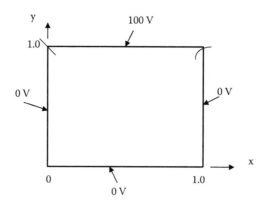

FIGURE 4.5
Figure for Example 4.2.

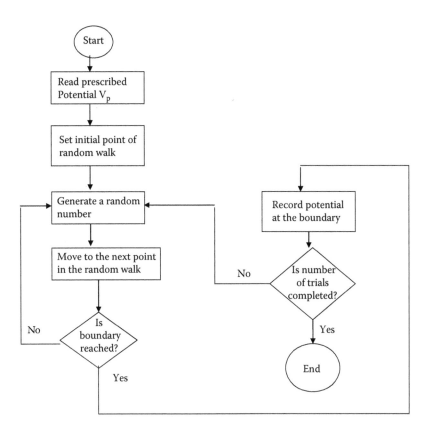

FIGURE 4.6
Flowchart for random walk of Example 4.2.

The exact solution obtained by the method of separation of variables is given in [9]

$$V(x_o, y_o) = \frac{400}{\pi} \sum_{n=1}^{\infty} \frac{\sin k\pi x \sinh k\pi y}{k \sinh k\pi}, \quad k = 2n+1 \tag{4.24}$$

Applying the fixed random MCM, the flowchart in Figure 4.6 was developed. Based on the flowchart, the program of Figure 4.7 was developed in MATLAB®. A built-in standard subroutine *rand* was used to generate random numbers U uniformly distributed between 0 and 1. The step size Δ was selected as 0.05. The results of the potential computation are listed in Table 4.1 for three different locations. The average number of random steps $\bar{m}$ taken to reach the boundary is also shown. It is observed from Table 4.1 that it takes a large number of random steps for small step size and that the error in MCM results can be less than 1%.

```
%   Using fixed random walk MCM to solve a potential

%   problem involving Laplace's equation; for Example 4.2


nrun = 2000;  % no. of runs, N

delta = 0.05; %  step size

A=1.0; B=1.0;

v1=0; v2=0; v3=100; v4=0;

xo=0.5;

yo=0.5;

io=xo/delta;   % location at which U is to be determined

jo=yo/delta;

imax=A/delta;

jmax=B/delta;

sum=0;

ms = 0;       % no. of steps before reaching the boundary

m1 = 0;       % no. of walks terminating at v1

m2 = 0;       % no. of walks terminating at v2

m3 = 0;       % no. of walks terminating at v3

m4 = 0;       % no. of walks terminating at v4

for k=1:nrun

  i=io;

  j=jo;

  while i<=imax & j<=jmax
```

FIGURE 4.7
MATLAB code for Example 4.2.

```
    ms = ms + 1;

r=rand; % random number between 0 and 1

if (r >= 0.0 & r <= 0.25)

  i=i+1;

end

if (r >= 0.25 & r <= 0.5)

  i=i-1;

end

if (r >= 0.5 & r <= 0.75)

  j=j+1;

end

if (r >= 0.75 & r <= 1.0)

  j=j-1;

end

  % check if (i,j) is on the boundary

  if(i == 0)

    sum=sum+ v4;

    m4 = m4 +1;

    break;

  end

  if(i == imax)

    sum=sum+ v2;

    m2 = m2 +1;
```

FIGURE 4.7
(Continued).

```
        break;
    end

            if(j == 0)
        sum=sum+ v1;
        m1 = m1 + 1;
            break;
    end

            if(j == jmax)
        sum=sum+ v3;
        m3 = m3 + 1;
            break;
    end

        end % while
    end
    v=sum/nrun
    m=ms/nrun
```

FIGURE 4.7
(Continued).

Rather than using Equation 4.12, an alternative approach to determine $V(x, y)$ is to calculate the probability of a random walk terminating at a grid point located on the boundary. The information is easily extracted from the program used for obtaining the results in Table 4.1. To illustrate the validity of this approach, the potential at (0.25, 0.75) was calculated. For $N = 1000$ random walks, the number of walks terminating at $x = 0$, $x = 1$, $y = 0$, and $y = 1$ are 461, 62, 66, and 411, respectively.
   Hence,

$$V(x_o, y_o) = \frac{1}{N} \sum_{k=1}^{m} N_k V_p(k) \qquad (4.25)$$

where m is the number of boundary sides with differential potential; N_k is the number of random walks terminal at side k of the boundary so that N_k/N

TABLE 4.1

Results of Example 4.2

x	y	N	$\bar{m}$	Monte Carlo Solution	Exact Solution
0.25	0.75	250	66.20	42.80	44.20
		500	69.65	41.80	
		750	74.19	41.60	
		1000	74.95	41.10	
		1250	74.67	42.48	
		1500	74.39	42.48	
		1750	74.08	42.67	
		2000	74.54	44.35	
0.5	0.5	250	118.62	21.60	25.0
		500	120.0	24.60	
		750	120.27	25.89	
		1000	120.92	25.80	
		1250	120.92	25.92	
		1500	120.78	25.27	
		1750	121.5	25.26	
		2000	121.74	25.10	
0.75	0.5	250	64.82	7.60	6.797
		500	68.52	6.60	
		750	68.56	6.93	
		1000	70.17	7.50	
		1250	72.12	8.00	
		1500	71.78	7.60	
		1750	72.40	7.43	
		2000	72.40	7.30	

is the probability that a random walk originating at (x, y) terminates at side k; and $V_p(k)$ is the potential there.
   Thus,

$$V(x_o, y_o) = \frac{461}{1000}(0) + \frac{62}{1000}(0) + \frac{66}{1000}(0) + \frac{411}{1000}(100) = 41.1 \qquad (4.26)$$

The statistical error in the simulation can be found. In this case, the potential on the boundary takes values 0 or $V_o = 100$, so that $V(x, y)$ has a binomial distribution with mean $V(x, y)$ and variance

$$\sigma^2 = \frac{V(x, y)[V_o - V(x, y)]}{N} \qquad (4.27)$$

At point (0.5, 0.5), for example, $N = 1000$ gives $\sigma = 1.384$ so that at 68% confidence interval, the error is $\delta = \sigma/\sqrt{N} = 0.04375$.

4.3 Solution of Poisson's Equation

The fixed random walk MCM can be used to solve Poisson's equation. To solve Poisson's equation,

$$\nabla^2 V = -g(x,y) \text{ in region R} \tag{4.28}$$

subject to

$$V = V_p \text{ on boundary B,} \tag{4.29}$$

we first obtain the finite difference representation for a square grid as

$$-g = [V(x+\Delta,y)+V(x-\Delta,y)+V(x,y+\Delta)+V(x,y-\Delta)-4V(x,y)]/\Delta^2$$

or

$$V(x,y)= p_{x+}V(x+\Delta,y)+p_{x-}V(x-\Delta,y)+p_{y+}V(x,y+\Delta)+p_{y-}V(x,y-\Delta)+\frac{\Delta^2 g}{4}$$

$$\tag{4.30}$$

where the transient probabilities remain as stated in Equation 4.10. The probabilistic interpretation of Equation 4.30 is similar to that of Equation 4.9. However, the term $\Delta^2 g/4$ in Equation 4.30 must be recorded at every step of the random walk. If m_i steps are required for the i-th random walk originating at (x,y) to reach the boundary, then one records

$$V_p(i)+\frac{\Delta^2}{4}\sum_{j=1}^{m_i-1} g(x_j,y_j) \tag{4.31}$$

at the end of that walk. Thus, the Monte Carlo estimate for $V(x_o, y_o)$ after N walks is

$$V(x_o,y_o)=\frac{1}{N}\sum_{i=1}^{N}V_p(i)+\frac{\Delta^2}{4N}\sum_{i=1}^{N}\left[\sum_{j=1}^{m_i-1} g(x_j,y_j)\right] \tag{4.32}$$

The Monte Carlo procedure for solving Poisson's equation may now be outlined.

To determine the potential at an arbitrary point (x_o, y_o) in the solution region, a series of random walks is constructed as follows:

1. Start each random walk at (x_o, y_o).
2. Generate a random number U, $0 < U < 1$, and move to a new point according to Equation 4.11 and record the value of $g(x, y)$ there.

3. If the new point is not on the boundary B, repeat step 2. If the new point is on B, terminate the random walk and record the prescribed potential $V_p(i)$ at the boundary point for the *i*-th random walk.
4. Repeat steps 1–3 for N random walks and then calculate

$$V(x_o, y_o) = \frac{1}{N} \sum_{i=1}^{N} V_p(i) + \frac{\Delta^2}{4N} \sum_{i=1}^{N} \left[\sum_{j=1}^{m_i - 1} g(x_j, y_j) \right] \qquad (4.33)$$

4.4 Solution of Axisymmetric Problems

In this section, we extend the random walk Monte Carlo procedures to potential problems with axisymmetric geometries [14].

Suppose the fixed random MCM is to be applied in solving Laplace's equation

$$\nabla^2 V = 0 \quad \text{in region R} \qquad (4.34)$$

subject to the Dirichlet boundary condition

$$V = V_p \quad \text{on boundary B} \qquad (4.35)$$

We begin by dividing the solution region R into a mesh and derive the finite difference equivalent of Equation 4.34. For $V = V(\rho, z)$, the problem is reduced to a two-dimensional one, and Equation 4.34 becomes

$$\frac{\partial^2 V}{\partial \rho^2} + \frac{1}{\rho} \frac{\partial V}{\partial \rho} + \frac{\partial^2 V}{\partial z^2} = 0 \qquad (4.36)$$

At any point (ρ, z) in R, $\rho \neq 0$, the finite difference approximation is

$$\frac{V(\rho + \Delta\rho, z) - 2V(\rho, z) + V(\rho - \Delta\rho, z)}{(\Delta\rho)^2} + \frac{1}{\rho} \frac{V(\rho + \Delta\rho, z) - V(\rho - \Delta\rho, z)}{2\Delta\rho}$$

$$+ \frac{V(\rho, z + \Delta z) - 2V(\rho, z) + V(\rho, z - \Delta z)}{(\Delta z)^2} \qquad (4.37)$$

where $\Delta\rho$ and Δz are the step sizes along ρ and z coordinates, as shown in the discretized typical solution of Figure 4.8. If, for ease of computation, we assume a square grid so that $\Delta\rho = \Delta z = \Delta$, Equation 4.37 becomes

$$V(\rho, z) = p_{\rho+} V(\rho + \Delta, z) + p_{\rho-} V(\rho - \Delta, y) + p_{z+} V(\rho, z + \Delta) + p_{z-} V(\rho, z - \Delta) \quad (4.38)$$

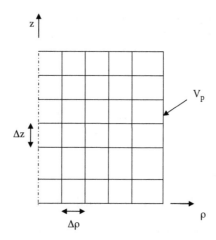

FIGURE 4.8
Typical axisymmetric solution region.

in which

$$p_{z+} = p_{z-} = \frac{1}{4} \tag{4.39a}$$

$$p_{\rho+} = \frac{1}{4} + \frac{\Delta}{8\rho} \tag{4.39b}$$

$$p_{\rho-} = \frac{1}{4} - \frac{\Delta}{8\rho} \tag{4.39c}$$

p_{z+}, p_{z-}, $p_{\rho+}$, and $p_{\rho-}$ are the transient probabilities. Note that

$$p_{z+} + p_{z-} + p_{\rho+} + p_{\rho-} = 1 \tag{4.40}$$

A probabilistic interpretation may be given to Equation 4.38. If a random-walking particle is momentarily at point (ρ, z), then p_{z+}, p_{z-}, $p_{\rho+}$, and $p_{\rho-}$ are the probabilities of stepping to points $(\rho, z + \Delta)$, $(\rho, z - \Delta)$, $(\rho + \Delta, z)$, and $(\rho - \Delta, z)$, respectively. Thus, by generating a random number U uniformly distributed between 0 and 1, we characterize the movement of the particle as follows:

$$\text{if } 0 < U < 0.25, \quad (\rho, z) \rightarrow (\rho, z + \Delta)$$

$$\text{if } 0.25 < U < 0.5, \quad (\rho, z) \rightarrow (\rho, z - \Delta)$$

$$\text{if } 0.5 < U < 0.75 + \frac{\Delta}{8\rho}, \quad (\rho, z) \rightarrow (\rho + \Delta, z)$$

$$\text{if } 0.75 + \frac{\Delta}{8\rho} < U < 1, \quad (\rho, z) \rightarrow (\rho - \Delta, z) \tag{4.41}$$

Equations 4.36–4.41 do not apply when $\rho = 0$. Because $V(\Delta, z) = V(-\Delta z)$, $\partial V/\partial \rho = 0$ at $\rho = 0$. Therefore, by L'Hospital's rule,

$$\lim_{\rho \to 0} \frac{1}{\rho} \frac{\partial V}{\partial \rho} = \frac{\partial^2 V}{\partial \rho^2} \tag{4.42}$$

Hence, at $\rho = 0$, Laplace's equation becomes

$$2 \frac{\partial^2 V}{\partial \rho^2} + \frac{\partial^2 V}{\partial z^2} = 0 \tag{4.43}$$

and the finite difference equivalent is

$$V(0, z) = p_{\rho+} V(\Delta, z) + p_{z+} V(0, z + \Delta) + p_{z-} V(0, z - \Delta) \tag{4.44}$$

where

$$p_{z+} = p_{z-} = \frac{1}{6}, \quad p_{z+} = \frac{4}{6}, \quad p_{z-} = 0 \tag{4.45}$$

Thus, at $\rho = 0$, the movement of the particle is decided by drawing a random number U and checking

$$\text{if } 0 < U < \frac{1}{6}, \quad (0, z) \to (0, z + \Delta)$$

$$\text{if } \frac{1}{6} < U < \frac{1}{3}, \quad (0, z) \to (0, z - \Delta)$$

$$\text{if } \frac{1}{3} < U < 1, \quad (0, z) \to (\Delta, z) \tag{4.46}$$

To estimate the potential at an arbitrary point (ρ_o, z_o) in the solution region, a series of random walks is constructed as follows:

1. Start each random walk at (ρ_o, z_o).
2. Generate a random number U, $0 < U < 1$, and move to a new point according to Equations 4.41 or 4.46.
3. If the new point is not on the boundary B, repeat step 2. If the new point is on B, terminate the random walk and record the prescribed potential $V_p(i)$ at the boundary point for the i-th random walk.
4. Repeat steps 1–3 for N random walks, and then calculate

$$V(\rho_o, z_o) = \frac{1}{N} \sum_{i=1}^{N} V_p(i) \tag{4.47}$$

The rate of convergence of the solution varies as $\sqrt{N}$, so that many random walks are necessary to guarantee accurate results.

To solve Poisson's equation

$$\nabla^2 V = -g \quad \text{in region R} \tag{4.48}$$

subject to

$$V = V_p \quad \text{on boundary B,} \tag{4.49}$$

we first obtain the finite difference form as

$$V(\rho, z) = p_{\rho+} V(\rho + \Delta, z) + p_{\rho-} V(\rho - \Delta, z) + p_{z+} V(\rho, z + \Delta)$$

$$+ p_{z-} V(\rho, z - \Delta) + \frac{\Delta^2 g}{4} \tag{4.50}$$

where the transient probabilities remain as defined in Equation 4.39. Thus, Equation 4.50 is subject to exactly the same probabilistic interpretation as Equation 4.38 except that the source term $\Delta^2 g/4$ must be recorded at each step of the random walk. If m_i steps are required to reach the boundary during the i-th random walk and g is assumed constant, instead of Equation 4.47, we now have

$$V(\rho_o, z_o) = \frac{1}{N} \sum_{i=1}^{N} V_p(i) + \frac{\Delta^2 g}{4} \frac{1}{N} \sum_{i=1}^{N} m_i \tag{4.51}$$

where m_i is the number of random steps taken to reach the boundary during the i-th walk.

It is also important to consider an extension of the MCM for the solution of problems with discrete homogeneities, that is, homogeneous media separated by interfaces. This requires a simple extension of what was said earlier for rectangular solution regions. For the z = constant interface, shown in Figure 4.9a, the boundary condition

$$D_{1n} = D_{2n} \quad \text{or} \quad \varepsilon_1 \partial V_1/\partial z = \varepsilon_2 \partial V_2/\partial z$$

leads to random walk probabilities

$$p_{z+} = \frac{\varepsilon_1}{2(\varepsilon_1 + \varepsilon_2)}$$

$$p_{z-} = \frac{\varepsilon_2}{2(\varepsilon_1 + \varepsilon_2)} \tag{4.52}$$

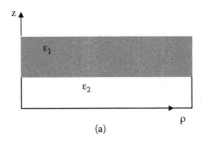

(a)

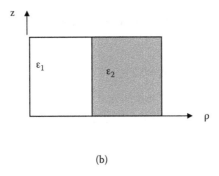

(b)

FIGURE 4.9
Interface between media of dielectric permittivities ε_1 and ε_2: (a) z = constant interface, (b) ρ = constant interface.

because $p_{\rho-}$ and $p_{\rho+}$ remain as in Equation 4.39. For ρ = constant interface, shown in Figure 4.9b, the boundary condition $\frac{\varepsilon_1}{\rho} \partial V_1/\partial \rho = \frac{\varepsilon_2}{\rho} \partial V_2/\partial \rho$ leads to

$$p_{\rho+} = \frac{\varepsilon_1}{2(\varepsilon_1 + \varepsilon_2)}\left(1 + \frac{\Delta}{2\rho}\right)$$

$$p_{\rho-} = \frac{\varepsilon_2}{2(\varepsilon_1 + \varepsilon_2)}\left(1 - \frac{\Delta}{2\rho}\right) \tag{4.53}$$

because p_{z-} and p_{z+} remain as in Equation 4.39.

Example 4.3
Find the potential distribution in a cylindrical earthed metal tank partially filled with charged liquid, assuming a uniform charge density. The problem

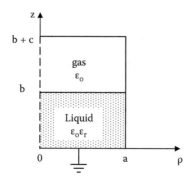

FIGURE 4.10
An earthed cylindrical tank partially filled with charged liquid; for Example 4.4.

is illustrated in Figure 4.10, where

$$a = b = c = 1.0 \text{ m}$$

$$\varepsilon_r = 2.0, \quad \rho_v = 10^{-5} \text{ C/m}^2$$

Region $z > b$ is gas, whereas $z < b$ is filled with a charged liquid such as hydrocarbons.

Solution

The analytic solution is presented in [15]

$$V_1 = 2\rho_v \sum_{n=1}^{\infty} \frac{1}{C_n K_n} [\cosh(\lambda_n b) - 1] \, J_o(\lambda_n \rho) \sinh[\lambda_n (b + c - z)], \quad z \geq b \qquad (4.54)$$

$$V_2 = \frac{2\rho_v}{\varepsilon_r} \sum_{n=1}^{\infty} \frac{J_0(\lambda_n \rho)}{C_n} \left(\begin{array}{l} \dfrac{\sinh(\lambda_n z)}{K_n} [\cosh(\lambda_n b) \cosh(\lambda_n c) \\ \\ + \varepsilon_r \sinh(\lambda_n b) \sinh(\lambda_n c) - \cosh(\lambda_n c)] + 1 - \cosh(\lambda_n z) \end{array} \right), \quad z \leq b$$

$$(4.55)$$

where subscripts 1 and 2 denote gas and liquid space, respectively,

$$C_n = \varepsilon_o a \lambda_n^3 J_1(\lambda_n a), \qquad (4.56)$$

$$K_n = \sinh(\lambda_n b) \cosh(\lambda_n c) + \varepsilon_r \cosh(\lambda_n b) \sinh(\lambda_n c), \qquad (4.57)$$

TABLE 4.2

Random Walk Probabilities and Source Term for Example 4.4

Location	$P_{\rho+}$	$P_{\rho-}$	P_{z+}	P_{z-}	g
$z > b, \rho \neq 0$	$\dfrac{1}{4} + \dfrac{\Delta}{8\rho}$	$\dfrac{1}{4} - \dfrac{\Delta}{8\rho}$	$1/4$	$1/4$	0
$z < b, \rho \neq 0$	$\dfrac{1}{4} + \dfrac{\Delta}{8\rho}$	$\dfrac{1}{4} - \dfrac{\Delta}{8\rho}$	$1/4$	$1/4$	$\dfrac{\rho_v \Delta^2}{4\varepsilon_2}$
$z = b, \rho \neq 0$	$\dfrac{1}{4} + \dfrac{\Delta}{8\rho}$	$\dfrac{1}{4} - \dfrac{\Delta}{8\rho}$	$\dfrac{\varepsilon_1}{2(\varepsilon_1 + \varepsilon_2)}$	$\dfrac{\varepsilon_2}{2(\varepsilon_1 + \varepsilon_2)}$	$\dfrac{\rho_v \Delta^2}{4(\varepsilon_1 + \varepsilon_2)}$
$z > b, \rho = 0$	$2/3$	0	$1/6$	$1/6$	0
$z < b, \rho = 0$	$2/3$	0	$1/6$	$1/6$	$\dfrac{\rho_v \Delta^2}{12\varepsilon_2}$
$z = b, \rho = 0$	$2/3$	0	$\dfrac{\varepsilon_1}{3(\varepsilon_1 + \varepsilon_2)}$	$\dfrac{\varepsilon_2}{3(\varepsilon_1 + \varepsilon_2)}$	$\dfrac{\rho_v \Delta^2}{12(\varepsilon_1 + \varepsilon_2)}$

and λ_n are the roots of $J_0(\lambda_n a) = 0$; and J_0 and J_1 are Bessel functions of order zero and one, respectively.

For the numerical solution, the random walk probabilities and the source terms at the boundary points and at the line of symmetry are derived in [16] and listed in Table 4.2. The potentials were calculated at four typical points using Equation 4.43. The computer program is in Figure 4.11. The results are displayed in Table 4.3, where δ is the error estimate. The step size $\Delta = 0.01$ was used in all calculations. In all Monte Carlo calculations, a built-in standard function *randu* in MATLAB was used to generate random numbers U uniformly distributed between 0 and 1.

The potential distribution for both the analytic and Monte Carlo results on the liquid–gas interface $z = b$ is shown in Figure 4.12 and along the z-axis in Figure 4.13. The solutions are close enough to verify the accuracy of the Monte Carlo calculations.

Example 4.4

Consider the potential problem shown in Figure 4.14. The problem is intractable by analytic methods, yet presents no real difficulties for a Monte Carlo treatment. This particular example is actually a prototype of an electrostatic particle focusing system that is employed in a recoil-mass time-of-flight spectrometer.

```
% This program solves an axisymmetric problem

% using the fixed randomwalk; for Example 4.4


delta=0.05;

a=1.0;

b=1.0;

c=1.0;

rhov=10^(-5);

eo=10^(-9)/(36*pi);

er1=1.0;

er2=2.0;

g=rhov*(delta)^2/eo;

rhoo=0.5; zo=0.5;  % initial point where v is to determined

sum=0.0;

sumg=0.0;

ms=0;

m=0; % no. of steps before reaching the boundary

nrun=1000;

for k=1:nrun

  z=zo;

  rho=rhoo;

  while ( rho >= 0 & rho <= a &  z >= 0 & z <= b+c )

    m=m+1;
```

FIGURE 4.11
MATLAB program for Example 4.4.

```
                    % calculate g; generate a random no. and move to the point

                    u = rand;

            if (z > b & rho ~= 0)

                    sumg = sumg + 0;

                    if(u>0 & u<=0.25)

                    z=z+delta;

                    end

                    if (u >0.25 & u <=0.50)

                    z=z-delta;

                    end

                    pp=0.25+ delta/(8* rho);

                    if(u >0.5 & u <= 0.5+pp)

                        rho=rho+delta;

                    end

                    if (u > 0.5+pp & u <= 1)

                        rho= rho- delta;

                    end

                end % end of segment 1


            if (z < b & rho ~= 0)

                    sumg = sumg + 1/(4*e r2);

                    if(u>0 & u<=0.25)

                    z=z+delta;
```

FIGURE 4.11
(Continued).

```
    end

    if (u >0.25 & u <=0.50)

    z=z-delta;

    end

    pp=0.25+ delta/(8*rho);

    if(u >0.5 & u <= 0.5+pp)

       rho=rho+delta;

    end

    if (u > 0.5+pp & u <= 1)

       rho= rho- delta;

    end

 end % end of segment 2

 if (z == b & rho ~= 0)

    sumg = sumg + 1/(4*( er1 + er2));

    pz1 = er1/(2*( er1 + er2));

    pz2 = er2/(2*( er1 + er2));

     if(u>0 & u<= pz1)

    z=z+delta;

    end

    if (u >pz1 & u <= pz1 + pz2)

    z=z-delta;

    end

    pp=0.25+ delta/(8*rho);
```

FIGURE 4.11
(Continued).

```
    if(u > 0.5 & u <= 0.5+pp)

       rho=rho+delta;

    end

    if (u > 0.5+pp & u <= 1)

       rho= rho- delta;

    end

end %  end of segment 3

if (z > b & rho == 0)

   sumg = sumg + 0;

   if(u>0 & u<=1/6)

   z=z+delta;

   end

   if (u >1/6 & u <= 1/3)

   z=z-delta;

   end

   if(u >1/3 & u <= 1.0)

      rho=rho+delta;

   end

end  %  end of segment 4

if (z < b & rho == 0)

   sumg = sumg + 1/(12*e r2);

   if(u>0 & u<=1/6)

   z=z+delta;
```

FIGURE 4.11
(Continued).

```
end

if (u >1/6 & u <= 1/3)

z=z-delta;

end

if(u >1/3 & u <= 1.0)

    rho=rho+delta;

end

end % end of segment 5

if (z == b & rho == 0)

    sumg = sumg + 1/(12*( er1 + er2));

    pz1 = er1/(3*( er1 + er2));

    pz2 = er2/(3*( er1 + er2));

    if(u>0 & u<= pz1)

    z=z+delta;

    end

    if (u >pz1 & u <= pz1 + pz2)

    z=z-delta;

    end

    if(u >1/3 & u <= 1)

        rho=rho+delta;

    end

end % end of segment 6

% check if the new point is on the boundary
```

FIGURE 4.11
(Continued).

```
    ms = ms +1;

    if (z == 0)

        sum = sum + 0;

        break;

    end

    if (z == b+c)

        sum = sum + 0;

        break;

    end

    if (rho == a)

    sum = sum + 0;

    break;

        end

    end % while

end % run

    % Calculate the potential

    v= sum/nrun + g*s umg/nrun
```

FIGURE 4.11
(Continued).

Solution

The finite difference solution to this problem is presented in [17]. The potentials were calculated at five typical points, and the results are compared with the finite difference results in Table 4.4. The step size $\Delta = 0.2$ was used in all Monte calculations. For the finite difference calculations, $\Delta = 0.25$ and $\Delta = 0.5$ gave nearly the same results for 500 iterations. Figures 4.15 and 4.16 show the potential distribution along $\rho = 5$ cm, $0 \leq z \leq 20$ cm and $z = 8$ cm, $1 \leq \rho \leq 20$ cm, respectively.

TABLE 4.3

Results of Example 4.3 (V is in kilovolts)

ρ	z	N	Monte Carlos Solution $(V \pm \delta)$	Exact Solution (V)
0.25	1.0	500	71.14 ±.016	71.12
		1000	70.8 ± 1.22	
		1500	69.74 ± 1.34	
		2000	69.45 ± 1.505	
0.75	1.0	500	34.02 ± 1.465	35.02
		1000	34.4 ± 0.8553	
		1500	34.68 ± 1.113	
		2000	34.24 ± 0.793	
0.5	1.5	500	12.4 ± 1.946	15.947
		1000	11.97 ± 0.6113	
		1500	11.91 ± 0.5207	
		2000	11.72 ± 0.4474	
0.5	0.5	500	60.14 ± 1.256	64.882
		1000	60.38 ± 2.226	
		1500	60.15 ± 1.51	
		2000	60.07 ± 1.662	

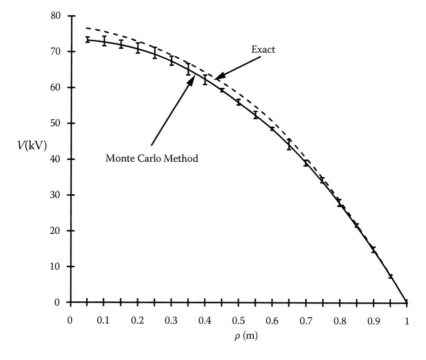

FIGURE 4.12

Potential distribution on liquid–gas interface; for Example 4.4.

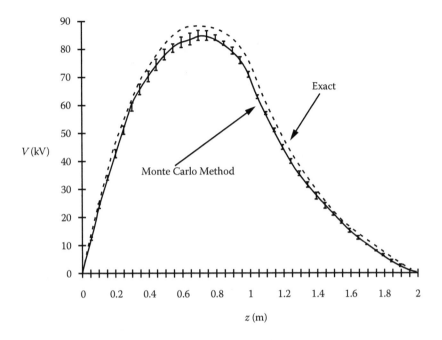

FIGURE 4.13
Potential distribution along the z-axis; for Example 4.4.

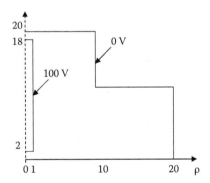

FIGURE 4.14
An electrostatic particle focusing system; for Example 4.4.

TABLE 4.4

Results of Example 4.4

ρ	z	N	Monte Carlos Solution $(V \pm \delta)$	Exact Solution (V)
5	18	500	10.52 ± 1.863	10.3467
		1000	10.80 ± 1.035	
		1500	10.83 ± 0.9135	
		2000	10.75 ± 0.6345	
5	10	500	26.36 ± 4.206	24.561
		1000	26.14 ± 2.26	
		1500	26.24 ± 1.605	
		2000	25.98 ± 1.777	
5	2	500	11.16 ± 1.463	10.404
		1000	11.48 ± 0.9908	
		1500	11.35 ± 1.167	
		2000	11.44 ± 0.8402	
10	2	500	2.56 ± 1.143	1.3026
		1000	2.66 ± 0.672	
		1500	2.493 ± 0.7621	
		2000	2.48 ± 0.5528	
15	2	500	0.6 ± 0.7239	0.1124
		1000	0.58 ± 0.3866	
		1500	0.5467 ± 0.3007	
		2000	0.49 ± 0.2648	

4.5 Summary

The MCM described in this chapter is called the *fixed random walk* type because the step size Δ is fixed and the steps of the walks are constrained to lie parallel to the coordinate axes. It provides a relatively straightforward means of solving potential problems with Dirichlet boundary conditions. The fixed random MCM is easy to program and gives results of sufficient accuracy for engineering purposes. One major disadvantage of this MCM is that it computes the potential at only one point at a time. The method may therefore be used to spot-check potentials obtained by other methods or for solving problems for which potentials at only a few points are required.

Another major drawback of the fixed random walk is that it is slow because a large number of steps per walk are required to reach the boundary due to the small, fixed step size. This drawback is fully removed by employing MCM with floating random walk, to be discussed in the next chapter.

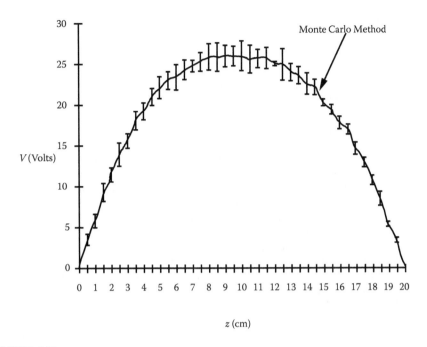

FIGURE 4.15
Potential distribution along ρ = 5 cm, 0 ≤ z ≤ 20 cm; for Example 4.4.

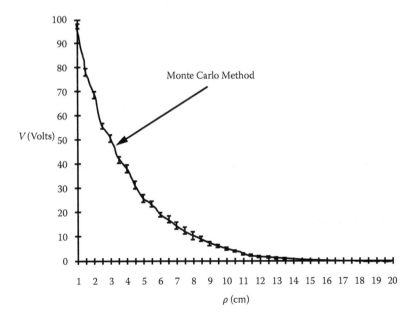

FIGURE 4.16
Potential distribution along z = 8 cm, 1 ≤ ρ ≤ 10 cm; for Example 4.4.

References

1. S. Kakutani, Two-dimensional Brownian motion harmonic functions, *Proc. Imp. Acad.* (Tokyo), vol. 20, 1944, pp. 706–714.
2. J. L. Doob, Interrelations between Brownian motion and potential theory, *Trans. Amer. Math. Soc.*, vol. 77, 86–121, July 1954.
3. J. L. Doob, A probability approach to the heat equation, *Trans. Amer. Math. Soc.*, vol. 80, 216–280, September 1955.
4. M. E. Muller, Some continuous Monte Carlo methods for the Dirichlet problem, *Ann. Math. Stat.*, vol. 27, 569–589, 1956.
5. G. M. Rayer, A Monte Carlo procedure for potential theory of problems, *IEEE Trans. Micro. Theo. Tech.*, Vol. 1–19, no. 10, Oct. 1971, p. 813–818.
6. M. N. O. Sadiku, Monte Carlo methods in an introductory electromagnetic course, *IEEE Trans. Educ.*, vol. 33, no. 1, 73–80, February 1990.
7. K. Binder and D. W. Heermann, *Monte Carlo Simulation in Statistical Physics: An Introduction*, Berlin: Springer-Verlag, 1992, pp. 8–14.
8. S. Caracciolo et al., Random walks with Short-range interaction and mean-field behavior, *J. Stat. Phys.*, vol. 77, no. 3/4, 519–544, 1994.
9. M. N. O. Sadiku, *Numerical Techniques in Electromagnetics*, 2nd ed., Boca Raton, FL: CRC Press, 2001, pp. 30–34, 137, 148–151.
10. P. G. Doyle and J. L. Snell, *Random Walks and Electric Networks*, Washington, D.C.: The Mathematical Association of America, 1984, pp. 3, 4.
11. Y. A. Shreider, *Method of Statistical Testing (Monte Carlo Method)*, Amsterdam: Elsevier, 1964, pp. 39–84. Another translation of the same Russian text: Y. A. Shreider, *The Monte Carlo Method (The Method of Statistical Trials)*, Oxford: Pergamon, 1966.
12. G. E. Zinsmeiter, Monte Carlo Methods as an aid in teaching heat conduction, *Bull. Mech. Engr. Educ.*, vol. 7, 77–86, 1968.
13. F. Sanchez-Quesada et al., Monte Carlo method for dielectric inhomogeneous problems, *Proc. IEE*, vol. 125, no. 12, December 1978, pp. 1400–1402.
14. M. N. O. Sadiku, Monte Carlo solution of axisymmetric potential Problems, *IEEE Trans. Indust. Appl.*, vol. 29, no. 6, 1042–1046, 1993.
15. K. Asano, Electrostatic potential and field in a cylindrical tank containing charged liquid, *Proc. IEE*, vol. 124, no. 12, December 1977, pp. 1277–1281.
16. M. N. O. Sadiku, S. O. Ajose, and F. Zhibao, Applying the exodus method to solve Poisson's equation, *IEEE Trans. Micro. Theo. Tech.*, vol. 42, no. 4, 661–666, April 1994.
17. M. DiStasia and W. C. McHarris, Electrostatic problems? Relax !, *Am. J. Phys.*, vol. 47, no. 5, 440–444, May 1979.

Problems

4.1. In Figure 4.1, use the fixed random walk Monte Carlo method to find $p(k)$ for $k = 2, 5$, and 9. Compare your result with the exact solution.

4.2. Given the one-dimensional differential equation

$$y'' = 0, \quad 0 \le x \le 1$$

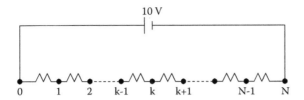

10 V

0   1   2   k-1   k   k+1   N-1   N

FIGURE 4.17
For Problem 4.4.

subject to $y(0) = 0$, $y(1) = 10$, use the fixed random walk Monte Carlo method to find $y(0.25)$ assuming

$\Delta x = 0.25$ and the following 20 random numbers:

0.1306, 0.422, 0.6597, 0.7905, 0.7695, 0.5106, 0.2961, 0.1428, 0.3666, 0.6543, 0.9975, 0.4866, 0.8239. 0.8722, 0.1330, 0.2296, 0.3582, 0.5872, 0.1134, 0.1404.

4.3. Consider N equal resistors connected in series as in Figure 4.17. By making $V(0) = 10$ V, find $V(k)$ using the fixed random walk Monte Carlo method for the following cases:

a. $N = 5, k = 2$

b. $N = 10, k = 7$

c. $N = 30, k = 11$

4.4. Consider the differential equation

$$\frac{\partial^2 W}{\partial x^2} + \frac{\partial^2 W}{\partial y^2} + \frac{k}{y}\frac{\partial W}{\partial y} = 0$$

where k = constant. By finding its finite difference equivalent, give a probabilistic interpretation to the equation.

4.5. Use the fixed random walk Monte Carlo procedure to determine the potential at points (2, 2), (3, 3), and (4, 4) in the problem shown in Figure 4.18a. By virtue of double symmetry, it is sufficient to consider a quarter of the solution region as shown in Figure 4.18b.

4.6. Use the fixed random walk Monte Carlo method to solve Laplace's equation in the triangular region $x \geq 0, y \geq 0, x + y \leq 1$ with the boundary condition $V(x, y) = x + y + 0.5$. Determine V at (0.4, 0.2), (0.35, 0.2), (0.4, 0.15), (0.45, 0.2), and (0.4, 0.25).

4.7. Apply Monte Carlo method to solve Laplace's equation in the three-dimensional region

$$|x| \leq 1, \quad |y| \leq 0.5, \quad |z| \leq 0.5$$

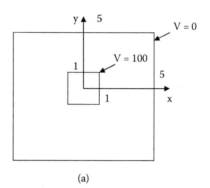

(a)

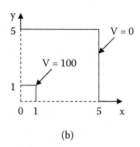

(b)

FIGURE 4.18
For Problem 4.5.

subject to the boundary condition

$$V(x, y, z) = x + y + z + 0.5.$$

Find the solution at (0.5, 0.1, 0.1).

4.8. Consider the potential system shown in Figure 4.19. Determine the potential at the center of the solution region. Take $\varepsilon_r = 2.25$.

4.9. For the potential system shown in Figure 4.20, find the potential at (0.25, 0.25), (0.5, 0.5), and (0.75, 0.5) assuming that

$$V_o = 100, \quad \varepsilon_1 = \varepsilon_o, \quad \varepsilon_2 = 3\varepsilon_0$$

$$a = b = 0.5, \quad h = w = 1.0$$

Note that at point (a,b)

$$p_{x+} = p_{y+} = \frac{\varepsilon_1}{3\varepsilon_1 + \varepsilon_2}, \quad p_{x-} = p_{y-} = \frac{(\varepsilon_1 + \varepsilon_2)}{2(3\varepsilon_1 + \varepsilon_2)}$$

4.10. In the solution region of Figure 4.21, $\rho_v = x(y-1)$ nC/m³. Find the potential at the center of the region using the fixed random walk Monte Carlo method.

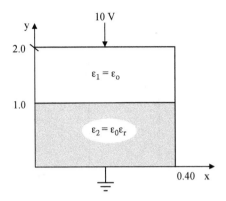

FIGURE 4.19
For Problem 4.8.

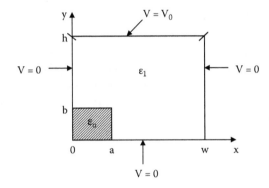

FIGURE 4.20
For Problem 4.9.

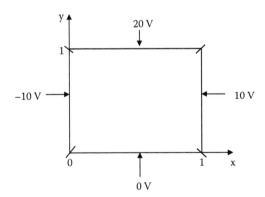

FIGURE 4.21
For Problem 4.10.

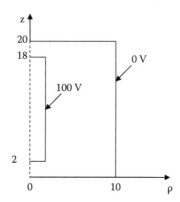

FIGURE 4.22
For Problem 4.12.

4.11. Derive the transient probabilities and source terms in Table 4.2 for the following cases: (a) $\rho = 0$, $z = b$, (b) $\rho = 0$, $z < b$.

4.12. Consider the finite cylindrical conductor held at $V = 100$ enclosed in a larger grounded cylinder as shown in Figure 4.22. Using the fixed random walk Monte Carlo technique, write a program to determine the potential at points $(\rho, z) = (2, 10)$, $(5, 10)$, $(8, 10)$, $(5, 2)$, and $(5, 18)$.

5

Floating Random Walk

Inexperience is what makes a young man do what an older man says is impossible.

—**Herbert V. Prochnow**

5.1 Introduction

There are two types of random walks in Monte Carlo methods (MCMs). The first and the more common method is the fixed random walk, where the step size is fixed. The second method is the floating random walk, where the step size changes from point to point. The second method takes less time because it takes fewer steps to reach the boundary and terminate a random walk. In the floating random walk MCM, neither the step nor the direction of the walk is fixed in advance. The quantities may be regarded as "floating" and hence the designation _floating random walk_. A floating random walk bypasses many intermediate steps of a fixed random walk in favor of a long jump. Fewer steps are needed to reach the boundary, and so computation is much more rapid than in the fixed random walk.

Unlike the deterministic numerical methods such as finite difference, finite elements, and moment methods, the floating random walk method is a nondeterministic (probabilistic or stochastic) numerical method employed in solving mathematical and physical problems. It is a means of treating mathematical problems by finding a probabilistic analog and then obtaining approximate answers to this analog by an experimental sampling procedure. The solution of a problem by this method is closer in spirit to physical experiments than to classical numerical techniques.

In this chapter, we will apply the floating random walk method to solve Laplace's and Poisson's equations for problems involving rectangular and axisymmetric solution regions that may or may not be inhomogeneous. As expected, using the floating random walk method, we can obtain an accurate solution in less time than by using the fixed random walk method.

5.2 Rectangular Solution Regions

5.2.1 Laplace's Equation

Suppose that the floating random walk is to be applied in solving Laplace's equation

$$\nabla^2 V = 0 \quad \text{in region R} \tag{5.1a}$$

subject to Dirichlet boundary condition:

$$V = V_p \quad \text{on boundary B} \tag{5.1b}$$

The mathematical basis of the floating random walk is the mean value theorem of potential theory in conjunction with Green function [1–3]. If S is a sphere of radius r centered at (x, y, z), which lies wholly within region R, then

$$V(x,y,z) = \frac{1}{4\pi a^2} \int_S V(r') dS' \tag{5.2}$$

Equation 5.2 can be given a probabilistic interpretation. The potential at the center of any sphere within R is equal to the average value of the potential taken over its surface. When the potential varies in two dimensions, $V(x, y)$ is given by

$$V(x,y) = \frac{1}{2\pi\rho} \oint_L V(\rho') dl' \tag{5.3}$$

where the integration is around a circle of radius ρ centered at (x,y). It can be shown that Equations 5.2 and 5.3 follow from Laplace's equation. Also, Equations 5.2 and 5.3 can be written as

$$V(x,y,z) = \int_0^1 \int_0^1 V(a,\theta,\phi) dF dT \tag{5.4}$$

$$V(x,y) = \int_0^1 V(a,\phi) dF \tag{5.5}$$

where

$$F = \frac{\phi}{2\pi}, \quad T = \frac{1}{2}(1 - \cos\theta) \tag{5.6}$$

and θ and ϕ are regular spherical coordinate variables. The functions F and T may be interpreted as the probability distributions corresponding to ϕ and θ, respectively. While $dF/d\phi$ = constant, $dT = \frac{1}{2}\sin\theta$; that is, all angles ϕ are equally probable, but the same is not true for θ.

The floating random walk MCM depends on the application of Equations 5.2 and 5.3 in a statistical sense. For a two-dimensional problem, suppose that a random-walking particle is at some point (x_j, y_j) after j steps in the

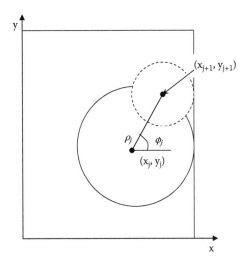

FIGURE 5.1
Configuration for floating random walk.

i-th walk. The next $(j+1)$-th step is taken as follows. First, a circle is constructed with center at (x_j, y_j) and radius ρ_j, which is equal to the shortest distance between point (x_j, y_j) and the boundary. The ϕ coordinate is generated as a random variable uniformly distributed over $(0, 2\pi)$, that is, $\phi = 2\pi U$, where $0 < U < 1$. Thus, the location of the random-walking particle after the $(j+1)$-th step is illustrated in Figure 5.1 and is given as

$$x_{j+1} = x_j + \rho_j \cos \phi_j \tag{5.7a}$$

$$y_{j+1} = y_j + \rho_j \sin \phi_j \tag{5.7b}$$

The next random step is made by constructing a circle centered at (x_{j+1}, y_{j+1}) and having a radius ρ_{j+1}, which is the shortest distance between (x_{j+1}, y_{j+1}) and the boundary. This procedure is repeated several times, and the walk is terminated when some prescribed small distance τ of the boundary is approached. The potential $V_p(i)$ at the end of this *i*-th walk is recorded as in fixed random walk MCM. The potential at (x_o, y_o) is eventually determined after N walks as

$$V(x_o, y_o) = \frac{1}{N} \sum_{i=1}^{N} V_p(i) \tag{5.8}$$

assuming that N is large.

The floating random walk MCM can be applied to a three-dimensional Laplace problem by proceeding along lines similar to those outlined here. A random-walking particle at (x_j, y_j, z_j) will step to a new location on the surface of a sphere whose radius r_j is equal to the shortest distance between point (x_j, y_j, z_j) and the boundary. The ϕ coordinate is selected as a random number U between 0 and 1, multiplied by 2π. The coordinate θ is determined by selecting another random number U between 0 and 1, and solving for $\theta = \cos^{-1}(1-2U)$, as in Chapter 2, Example 2.3. Thus, the location of the particle after its $(j+1)$-th step in the i-th random walk is

$$x_{j+1} = x_j + r_j \cos\phi_j \sin\theta_j \qquad (5.9a)$$

$$y_{j+1} = y_j + r_j \sin\phi_j \sin\theta_j \qquad (5.9b)$$

$$z_{j+1} = z_j + r_j \cos\theta_j \qquad (5.9c)$$

Finally, we apply Equation 5.8.

5.2.2 Poisson's Equation

Suppose that the floating random walk MCM is to be applied in solving Poisson's equation

$$\nabla^2 V = -g(x,y) = -\frac{\rho_v}{\varepsilon} \quad \text{in region R} \qquad (5.10a)$$

subject to the Dirichlet boundary condition

$$V = V_p \quad \text{on boundary B} \qquad (5.10b)$$

we apply the mean value theorem. If S is a sphere of radius a, centered at (x, y, z), which lies wholly within region R, then the solution to Equation 5.10 is

$$V(x,y,z) = \frac{1}{4\pi a^2}\int_S V(r')dS' + \frac{1}{4\pi}\int_v \frac{g(r')}{|r-r'|}dv' \qquad (5.11)$$

This can be given a probabilistic interpretation. The value $V(x,y,z)$ at the center of the sphere is the average of g with respect to Green's function plus the uniform average of g with respect to Green's function plus the uniform average of the boundary values on the surface of the sphere. When the potential varies in two dimensions, $V(x,y)$ is given by

$$V(x,y,z) = \frac{1}{2\pi a}\oint_L V(\rho')d\rho' + \frac{1}{2\pi}\int_S g(\rho')\ln|\rho-\rho'|dS' \qquad (5.12)$$

where the integration is around a circle of radius a centered at (x, y). Alternatives to Equations 5.11 and 5.12 are

$$V(x,y,z) = \int_0^1\int_0^1 [V(a,\theta,\phi) + a^2 g/6]dFdT \qquad (5.13)$$

$$V(x,y) = \int_0^1 [V(a,\phi) + a^2 g/4]dF \qquad (5.14)$$

where F and T remain as defined in Equation 5.6.

The floating random walk MCM depends on the application of Equations 5.11 and 5.12 in a statistical sense. For a two-dimensional problem, the location of the random-walking particle after the j-th step is given by

$$x_{j+1} = x_j + \rho_j \cos \phi_j \qquad\qquad (5.15a)$$

$$y_{j+1} = y_j + \rho_j \sin \phi_j \qquad\qquad (5.15b)$$

The value of $g(x, y) \bullet \rho^2$ at this new location is recorded. The next random step is executed by constructing a circle centered at (x_{j+1}, y_{j+1}) and having a radius ρ_{j+1}, which is the shortest distance to the boundary. This procedure is repeated several times, and the walk is terminated when some prescribed small distance τ from the boundary is approached. The potential $V_p(i)$ at the closest boundary point at the end of the i-th random walk is recorded as in fixed random walk MCM. The potential at point (x_o, y_o) is eventually determined after N walks using

$$V(x_o, y_o) = \frac{1}{N} \sum_{i=1}^{N} V_p(i) + \frac{1}{4N} \sum_{i=1}^{N} \left[\sum_{j=1}^{m_i} g(x_j, y_j)\rho_j^2 \right] \qquad\qquad (5.16)$$

where m_i is the number of random steps required to reach the boundary in the i-th walk.

The floating random walk MCM can be applied to a three-dimensional problem involving Poisson's equation by proceeding along lines similar to those outlined here. The location of the particle after $(j + 1)$-th step in the i-th random walk is determined using Equation 5.9, and the value of $g(x, y, z) \bullet r^2$ at this new location is recorded. Eventually, the potential at (x_o, y_o, z_o) is determined as

$$V(x_o, y_o, z_o) = \frac{1}{N} \sum_{i=1}^{N} V_p(i) + \frac{1}{6N} \sum_{i=1}^{N} \left[\sum_{j=1}^{m_i} g(x_j, y_j, z_j)r_j^2 \right] \qquad\qquad (5.17)$$

The floating random walk described here has a varying step size; that is, the step size changes from point to point in the random walk [4]. The step size could be fixed (as in fixed random walk), in which case it is called *fixed-radius floating random* walk [5].

Example 5.1

Use the floating random walk MCM to determine the potential at points (1.5, 0.5), (1.0, 1.5), and (1.5, 2.0) in the two-dimensional region of Figure 5.2.

Solution

To apply the floating random walk, we use the flowchart in Figure 5.3 except that we apply Equation 5.7 instead of Equation 4.11 (Chapter 4) at every step

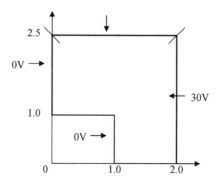

FIGURE 5.2
Figure for Example 5.1.

in each random walk. A computer program based on the modified flowchart was developed, as shown in Figure 5.3. The shortest distance ρ from (x, y) to the boundary was found by dividing the solution region in Figure 5.2 into three regions and checking

$$\text{if } \{(x, y): 1 < x < 2, 0 < y < 1\}, \rho = \text{minimum } \{x - 1, 2 - x, y\}$$

$$\text{if } \{(x, y): 0 < x < 1, 1 < y < 2.5\}, \rho = \text{minimum } \{x, y - 1, 2.5 - y\}$$

$$\text{if } \{(x, y): 1 < x < 2, 1 < y < 2.5\}, \rho = \text{minimum}, \left\{2 - x, 2.5 - y, \sqrt{(x-1)^2 + (y-1)^2}\right\}$$

A prescribed tolerance $\tau = 0.005$ was selected so that if the distance between a new point in the random walk and the boundary is less than τ, it is assumed that the boundary is reached and the potential at the closest boundary point is recorded.

Table 5.1 presents the Monte Carlo results with the average number of random steps $\bar{m}$. It should be noted that it takes fewer walks to reach the boundary in floating random walk than in fixed random walk. Because no analytic solution exists, we compare Monte Carlo results with those obtained using finite difference method with $\Delta = 0.05$ and 500 iterations. As evident from Table 5.1, the Monte Carlo results agree well with the finite difference results even with 1000 walks.

Unlike the program in Chapter 4, Figure 4.7, where the error estimates are not provided for the sake of clarity and simplicity, the program in Figure 5.3 incorporates evaluation of error estimates in the Monte Carlo calculations. Using Equation 2.46, (Chapter 2) the error is calculated as

$$\delta = \frac{S t_{\alpha/2; n-1}}{\sqrt{n}}$$

In the program in Figure 5.3, the number of trials n is taken as 5 so that $t_{\alpha; n-1} = 2.766$. The sample variance S is calculated using Equation 2.33

```
% Using floating random walk MCM
% to solve Laplace's equation; for Example 5.1


nrun = 500;
ntrials = 5;    % no. of trials
tol=0.005;      % tolerance
xo = 1.5;       % location at which potential
yo = 0.5;       % is to be determined
for n=1:ntrials
sum = 0.0;
m=0;    % no. of steps before reaching the boundary
for k=1:nrun
   x=xo;
   y=yo;
  while x >=0 & x <=2  & y >=0 & y<=2.5
     u=rand;  % generate a random no. and move to the next point
     phi=2.0*pi*u;
     % find the shortest distance r;
     rc = sqrt( (x-1)^2 + (y-1)^2 );
     if (x >1 & x < 2 & y >0 & y < 1)
       rx = min(x-1,2-x);
       r = min(rx,y);
     end
     if (x >0 & x < 1 & y >1 & y < 2.5)
       rx = min(x,y-1);
       r = min(rx,2.5-y);
     end
     if (x >1 & x < 2 & y >1 & y < 2.5)
       rx = min(2-x,2.5-y);
```

(a)

FIGURE 5.3
MATLAB code for Example 5.1.

```
      r = min(rx,rc);

   end

   x=x+r*cos(phi);

   y=y+r*sin(phi);

   m=m+1;

   % check if (x,y) is on the boundary

   if ( x < (1+tol) & y < (1+tol) ) % corner

       sum=sum +0;

       break;

   end

     if ( x >= (2-tol) )

       sum=sum +30;

       break;

   end

   if ( y >= (2.5-tol) )

       sum=sum +20;

       break;

   end

   if ( x < tol & y >1 & y < 2.5)

       sum=sum +0;

       break;

   end

   if (   x > tol & x < (1-tol) & y <= (1+tol)   )

       sum=sum +0;

       break;

   end

   if ( x < (1+tol) & y >= tol & y <= 1 )

       sum=sum +0;

       break;
```

(b)

FIGURE 5.3
(Continued).

```
                    end

              if ( x >= (1+tol) & x <= (2.5-tol)   & y < tol )

                    sum=sum +0;

                    break;

              end

         end %while

  end % nrun

  vv(n)=sum/nrun;

  steps(n)=m/nrun;

  end % ntrials

  % find the mean value of V and mean no. of steps

  sum1 = 0.0;

  sum2 = 0.0;

  for n=1:ntrials

       sum1 = sum1 + vv(n);

       sum2 = sum2 + steps(n);

  end

  vmean=sum1/ntrials

  stepm=sum2/ntrials

  % calculate error

  sum3=0.0;

  for n=1:ntrials

       sum3 = sum3 + ( vv(n)  - vmean )^2;

  end

  std=sqrt( sum3/(ntrials-1) );

  error = std*2.776/sqrt(ntrials)
```

(c)

FIGURE 5.3
(Continued).

TABLE 5.1

Results of Example 5.1

x	y	N	$\bar{m}$	Monte Carlo Solution $(V \pm \delta)$	Finite Difference Solution (V)
1.5	0.5	250	6.738	11.52 ± 0.8973	11.44
		500	6.668	11.80 ± 0.9378	
		750	6.535	11.83 ± 0.4092	
		1000	6.476	11.82 ± 0.6205	
		1250	6.483	11.85 ± 0.6683	
		1500	6.465	11.72 ± 0.7973	
		1750	6.468	11.70 ± 0.6894	
		2000	6.460	11.55 ± 0.5956	
1.0	1.5	250	8.902	10.74 ± 0.8365	10.44
		500	8.984	10.82 ± 0.3709	
		750	8.937	10.75 ± 0.5032	
		1000	8.928	10.90 ± 0.7231	
		1250	8.836	10.84 ± 0.7255	
		1500	8.791	10.93 ± 0.5983	
		1750	8.788	10.87 ± 0.3646	
		2000	8.811	10.84 ± 0.3646	
1.5	2.0	250	7.242	21.66 ± 0.7509	21.07
		500	7.278	21.57 ± 0.5162	
		750	7.278	21.53 ± 0.3505	
		1000	7.316	21.53 ± 0.2601	
		1250	7.322	21.53 ± 0.3298	
		1500	7.348	21.51 ± 0.3083	
		1750	7.372	21.55 ± 0.2592	
		2000	7.371	21.45 ± 0.2521	

Note: $\bar{m}$ = average number of random steps.

(Chapter 2). The values of δ are also listed in Table 5.1. Note that unlike in Chapter 3, Table 3.1, where $\bar{m}$ and V are the mean values after N walks, $\bar{m}$ and V in Table 5.1 are the mean values of n trials, each of which involves N walks, that is, the "mean of the mean" values. Hence, the results in Table 5.1 should be regarded as more accurate than those in Table 4.1 (Chapter 4).

Example 5.2

Consider Poisson's equation

$$\nabla^2 V = g(x,y) = -36\pi x(y-1), \quad 0 \le x \le 1, 0 \le y \le 1$$

subject to boundary conditions

$$V(x,0) = 0, \quad V(x,1) = 20, \quad V(0,y) = -10, \quad V(1,y) = 10$$

TABLE 5.2

Results of Example 5.2

x	y	Exact Solution (V)	Fixed Random Walk $(V \pm \delta)$	Floating Random Walk $(V \pm \delta)$
0.25	0.25	−3.425	−3.402 ± 0.106	−3.430 ± 0.0630
0.75	0.25	2.902	2.908 ± 0.1073	2.943 ± 0.0804
0.5	0.5	2.913	2.993 ± 0.1127	2.995 ± 0.3275
0.25	0.75	5.277	5.242 ± 0.2394	5.241 ± 0.1285
0.75	0.75	11.131	11.146 ± 0.1369	11.208 ± 0.0714

Solution

The analytic solution to this problem using the series expansion technique is given in [1, pp. 156–161], in which the finite difference solution is also available.

The potentials were calculated at five typical points using both fixed and floating random walks. For the fixed random walk, $\Delta = 0.05$ and $N = 10{,}000$ walks were used. The same number of walks was used for the floating random walk. The tolerance $\tau = 0.005$ was used for all floating random walks. The results are displayed in Table 5.2. In the table, δ is the error estimate, which is obtained by repeating each calculation five times and using Equation 2.46 (Chapter 2).

5.3 Axisymmetric Solution Regions

We now apply what we have covered in Section 5.2 to the case when the solution region is axisymmetric.

5.3.1 Laplace's Equation

Suppose that the floating random walk method is to be applied in solving Laplace's equation subject to the Dirichlet boundary condition:

$$\nabla^2 V(X) = 0 \quad X \in \Omega$$

$$V(X) = V_p \quad X = \partial\Omega \tag{5.18}$$

Let a particle, initially positioned within a closed surface S_b at X_0, undergo Brownian motion until it reaches boundary S_b. Kakutani [6] showed that the expected value of the potential at the position where the particle reaches S_b, $(E[V(S_b)])$, was the solution of the Dirichlet problem for potential at X_0:

$$V(X_0) = E[V(S_b)] \tag{5.19}$$

$$E[V(S_b)] = \lim_{n \to \infty} \sum_{i=1}^{n} \frac{V_p(i)}{n} \tag{5.20}$$

where $V_p(i)$ is the potential at the position at which the aforementioned particle reaches S_b in the experiment and the step size for each random walk is "floating."

5.3.2 Poisson's Equation

Consider Poisson's equation with the Dirichlet boundary condition:

$$\nabla^2 V(X) = -g(X) \quad X \in \Omega$$

$$V(X) = f(X) \quad X \in \partial\Omega \tag{5.21}$$

We assume that the boundary $\partial\Omega$ is sufficiently smooth so as to ensure the existence of a unique solution $V(X)$ that is twice continuously differentiable in Ω and continuous on $\Omega \cup \partial\Omega$. Also, we assume that q is continuous in $\Omega \cup \partial\Omega$ and that the domain Ω consists of a finite union of bounded convex regions. In addition, we suppose that the Green's function $G(X, X_0)$ for the Dirichlet problem on Ω exists and that it is continuously differentiable for $X \in \Omega \cup \partial\Omega$. It can be shown that the solution to Equation 5.21 has integral representation [1, 7]:

$$V(X_0) = \int_{B(X_0)} g(X)G(X, X_0)dX + \frac{1}{|S(X_0)|} \int_{S(X_0)} V(X)d\sigma \tag{5.22}$$

The probabilistic interpretation of Equation 5.22 is given as follows: the value $V(X_0)$ at the center of the sphere is the average of $g(X)$ with respect to the Green's function plus the uniform average of the boundary value on the surface of the ball. When the boundary values on the sphere are known, $V(X_0)$ could be interpreted as randomly sampling $q(X)$ inside the ball (to estimate the first integral) and uniformly selecting $V(X)$ on the sphere (to estimate the second integral). Averaging over a large number of trials and summing would give an estimate of $V(X_0)$.

Alternatives to Equation 5.22 in two and three dimensions, respectively, are given as

$$V(X_0) = \int_0^1 \int_0^1 \left[V(a, \theta, \phi) + \frac{a^2 q(X)}{6} \right] dFdT \tag{5.23}$$

$$V(X_0) = \int_0^1 \left[V(a, \phi) + \frac{a^2 q(X)}{4} \right] dF \tag{5.24}$$

where

$$F = \frac{\phi}{2\pi}, \quad T = \frac{1}{2}(1 - \cos\theta) \tag{5.25}$$

and a, θ, and ϕ are regular spherical coordinate variables. For axisymmetric problems, we only discuss floating random walk for two-dimensional solution regions.

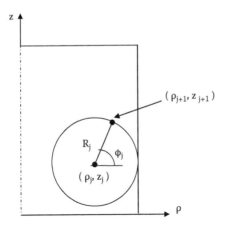

FIGURE 5.4
Obtaining the next point in a floating random walk.

5.3.3 Homogeneous Media

The floating random walk method depends on the statistical application of Equations 5.23 and 5.24. For a two-dimensional problem, we suppose that a random walking particle is at some point (ρ_j, z_j) after the j-th step in the i-th walk (see Figure 5.4). The next $(j + 1)$-th step is taken as follows. First, a circle is constructed with center at (ρ_j, z_j) and radius R_j, which is equal to the shortest distance between (ρ_j, z_j) and the boundary. The ϕ_j coordinate is generated as a random variable uniformly distributed over each subregion $2\pi/M$ (M is the number of subregions). Thus, the location of the random-walking particle after the $(j + 1)$-th step is illustrated in Figure 5.4 and given as follows:

$$\rho_{j+1} = \rho_j + R_j \cos \phi_j$$

$$z_{j+1} = z_j + R_j \sin \phi_j \tag{5.26}$$

This procedure is repeated until the walk is terminated when some prescribed small distance τ from the boundary is approached. That is the end of one random walk. We conduct N random walks and record the potential at the end of each random walk. When N is large enough, the potential at (ρ_0, z_0) is eventually determined using

$$V(\rho_0, z_0) = \frac{1}{N} \sum_{i=1}^{N} V_p(i) + \frac{1}{4N} \sum_{i=1}^{N} \left[\sum_{j=1}^{m_i} q(\rho_j, z_j) R_j^2 \right] \tag{5.27}$$

where $V_p(i)$ is the potential at the closest boundary point at the end of the i-th random walk, and m_i is the number of steps taken to reach the boundary during the i-th random walk.

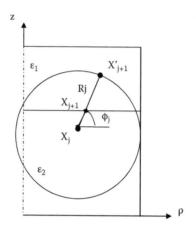

FIGURE 5.5
Interface between two media with dielectric permittivities ε_1 and ε_2.

5.3.4 Inhomogeneous Media

For the case of two media of dielectric permittivities ε_1 and ε_2 (Figure 5.5), the boundary condition $(D_{1n} = D_{2n})$ and Gauss's law for the electric field lead to random walk transient probabilities on the interface [8]:

$$P_{z+} = \frac{\varepsilon_1}{2(\varepsilon_1 + \varepsilon_2)}$$

$$P_{z-} = \frac{\varepsilon_2}{2(\varepsilon_1 + \varepsilon_2)} \tag{5.28}$$

Suppose that a random-walking particle is at X_j after the j-th step. After the $(j + 1)$-th step, the particle moves to $X_{j+1'}$ and crosses the interface between two media. Because there are differential transient probabilities to media ε_1 and ε_2 on the interface, the random-walking particle must move from $X_{j+1'}$ to X_{j+1} along the line $X_j\,X_{j+1'}$, then we use Equation 5.28 to determine the transfer from interface to ε_1 or ε_2 medium. The interior point contribution on the interface to the potential is as follows:

$$V = \frac{1}{8}[q_1(\rho_j, z_j) + q_2(\rho_j, z_j)]R_j^2 \tag{5.29}$$

where

$$q_1(\rho_j, z_j) = \lim_{\rho \to \rho_j, z \to z_j} q_1(\rho, z)$$

$$q_2(\rho_j, z_j) = \lim_{\rho \to \rho_j, z \to z_j} q_2(\rho, z) \tag{5.30}$$

5.3.5 The Computing Procedure

The computing procedure for solving an axisymmetric problem is summarized as follows:

1. Divide 2π region into 8 ($\pi/4$ each), 16 ($\pi/8$ each), 32($\pi/16$ each), or M subregions. The larger the value of M, the higher the accuracy.
2. Begin a random walk at (ρ_0, z_0), where the potential is desired.
3. Determine the shortest distance R_j to the boundary and calculate the following:

$$\rho_{j+1} = \rho_j + R_j \cos \phi_j$$

$$z_{j+1} = z_j + R_j \sin \phi_j \tag{5.31}$$

   If point (ρ_j, z_j) is an interior point, then ϕ_j is $0 \le \phi_j \le 2\pi$; if point (ρ_j, z_j) is on the z-axis (symmetric axis), then ϕ_j is $-\pi/2 \le \phi_j \le \pi/2$.
4. If the next step crosses the interface between two media, the new point should be X_{j+1} (shown in Figure 5.5). If the new point is on the interface between two media, we first determine the transient probabilities using Equation 5.28 and then locate the new position X_{j+1}, and ϕ_j is $0 \le \phi_j \le \pi$ or $\pi \le \phi_j \le 2\pi$ depending on the transient probabilities in Equation 5.28. If $\rho_{j+1} < 0$, it means that new point falls outside the solution region in Figure 5.5 and we set $\rho_{j+1} = -\rho_{j+1}$ because of the symmetry.
5. When the random walk falls within a τ-region from boundary (where τ is a preassigned tolerance value), we stop the i-th random walk and repeat steps 2 to 4 for a very large number N of walks. Finally, we use Equation 5.27 to calculate the potential $V(\rho_0, z_0)$.

We shall illustrate the application of the present method by means of two numerical examples that have analytic solutions so that the accuracy and validity of the floating random walk method can be checked. The first example is an axisymmetric problem that involves a homogeneous solution region, whereas the second example involves an inhomogeneous solution region.

Example 5.3

Consider the solution of an axisymmetric problem involving Laplace's equation

$$\nabla^2 V = \frac{1}{\rho} \frac{\partial}{\partial \rho}\left(\rho \frac{\partial V}{\partial \rho}\right) + \frac{\partial^2 V}{\partial z^2} = 0 \tag{5.32}$$

subject to

$$V(\rho, 0) = V(\rho, L) = 0$$

$$V(a, z) = V_o \tag{5.33}$$

Solution

The exact solution is

$$V(\rho, z) = \frac{4V_o}{\pi} \sum_{n=\text{odd}}^{\infty} \frac{I_0(n\pi\rho/L)}{nI_0(n\pi a/L)} sin(n\pi z/L) \tag{5.34}$$

where I_0 is the modified Bessel function of the first kind and order zero. For numerical calculation, take $L = 2a = 1$ m and $V_o = 100$ V. Table 5.3 compares the floating random walk solution with the exact solution. For the floating random walk,

$$R = \text{minimum } (a - \rho, z, L - z)$$

A prescribed tolerance $\tau = 0.05$ was selected so that if the distance between a new point in the random walk and the boundary is less than τ, it is assumed

TABLE 5.3

Comparison of the Floating Random Walk Solution with the Exact Solution

(ρ, z)	Exact Solution	Floating Random Walk	
(0, 0.25)	53.75	50.9	$N = 1000$
		52.54	$N = 5000$
		52.96	$N = 10000$
		52.28	$N = 15000$
(0, 0.5)	72.13	68.5	$N = 1000$
		69.56	$N = 5000$
		69.79	$N = 10000$
		69.97	$N = 15000$
(0, 0.75)	53.75	52.3	$N = 1000$
		53.16	$N = 5000$
		52.74	$N = 10000$
		52.29	$N = 15000$
(0.125, 0.5)	75.33	71.4	$N = 1000$
		72.68	$N = 5000$
		72.21	$N = 10000$
		72.34	$N = 15000$
(0.25, 0.5)	80.65	78.6	$N = 1000$
		77.96	$N = 5000$
		78.33	$N = 10000$
		78.51	$N = 15000$

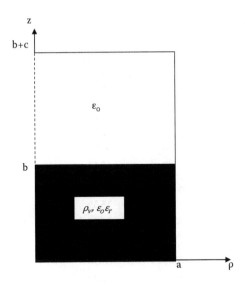

FIGURE 5.6
Partially filled earthed cylindrical tank.

that the boundary is reached and the potential at the closest boundary point is recorded.

Example 5.4
Consider an earthed cylindrical tank partially filled with charged liquid (Figure 5.6), where

$$a = b = 1.0\text{m} \qquad c = 2.0\text{m}$$

$$\varepsilon_r = 2.0 \qquad \rho_v = 10^{-5}\text{C/m}^3$$

Region $z > b$ is vacuum, whereas $z < b$ is filled with a charged liquid.

Solution
The exact solution is [9]

$$V_1 = 2\rho_v \sum_{n=1}^{\infty} \frac{1}{C_n K_n}[\cosh(\lambda_n b) - 1]J_0(\lambda_n \rho)\sinh[\lambda_n(b+c-z)], \quad z \geq b \quad (5.35)$$

$$V_2 = \frac{2\rho_v}{\varepsilon_r} \sum_{n=1}^{\infty} \frac{J_0(\lambda_n \rho)}{C_n} \left(\begin{array}{l} \dfrac{\sinh(\lambda_n z)}{K_n}[\cosh(\lambda_n b)\cosh(\lambda_n c) \\ +\varepsilon_r \sinh(\lambda_n b)\sinh(\lambda_n c) - \cosh(\lambda_n c)] + 1 - \cosh(\lambda_n z) \end{array} \right), z \leq b$$

$$(5.36)$$

TABLE 5.4

Results of Example 5.5

ρ	z	N	Floating Random Walk (kV)	Exact Solution (kV)
0.5	0.5	15000	63.30	63.88
		10000	65.70	
		5000	66.59	
		1000	67.95	
0.5	1.5	15000	15.13	15.94
		10000	15.07	
		5000	15.87	
		1000	15.58	
0.75	1.0	15000	37.70	35.02
		10000	37.83	
		5000	37.82	
		1000	38.56	
0.25	1.0	15000	72.97	71.12
		10000	72.62	
		5000	72.52	
		1000	72.56	

where subscripts 1 and 2 denote gas and liquid spaces, respectively,

$$C_n = \varepsilon_o a \lambda_n^3 J_1(\lambda_n a) \tag{5.37}$$

$$K_n = \sinh(\lambda_n b)\cosh(\lambda_n c) + \varepsilon_r \cosh(\lambda_n b)\sinh(\lambda_n c) \tag{5.38}$$

and λ_n are the roots of $J_o(\lambda_n a) = 0$.

The potentials were calculated at four typical points using the floating random walk Monte Carlo method. The results are compared with the exact solution in Table 5.4.

In general, we note two main factors that influence the accuracy of floating random walk. One factor is that the boundary has been replaced by a τ-region and we take the value of the τ-region to be the same as the boundary value. The second is that 2π region is divided into a limited number of subregions. When τ is decreased, so is the error, but the computational time is increased. When a large number (M) of subregions is taken, the accuracy increases. We need to choose a reasonable τ and M while keeping the error as small as possible. In the two examples, we take $\tau = 0.05$ and $M = 16$.

5.4 Summary

The floating random walk Monte Carlo method has been shown to be applicable for solving the potential problems of Laplace's and Poisson's field distribution with Dirichlet's condition for rectangular and axisymmetric solution regions.

It has also been shown that the method can handle inhomogeneous solution regions. The method provides an accurate solution quickly compared to fixed random walk. The accuracy of the floating random walk method is dependent on a preset τ-region. The smaller the value of τ, the higher the accuracy, but the longer the computational time. The probabilistic scheme can be very useful for solving a problem that has a severe gradient near the boundary point. The major limitation of the method is that it calculates the potential at only one point at a time. This limitation is common with other classical Monte Carlo methods. To overcome this limitation, we can use Green's function in the floating random walk [10,11] or use Markov chain, which will be discussed in Chapter 8.

References

1. M. N. O. Sadiku, *Numerical Techniques in Electromagnetics*, 3rd ed., Boca Raton: CRC Press, 2009, Chap. 8.
2. A. Haji-Sheikh and E. M. Sparrow, The floating random walk and its application to Monte Carlo solutions of heat equations, *SIAM J. Appl. Math.*, vol. 14, no. 2, 370–389, March 1966.
3. J. N. Jere and Y. L. Le Coz, An improved floating-random walk algorithm for solving the multi-dielectric Dirichlet problem, *IEEE Trans. Microwave Theory Tech.*, vol. 41, no. 3, 325–329, February 1993.
4. M. N. O. Sadiku and R. C. Garcia, Monte Carlo floating walk solution of Poisson's equation, *Proc. of IEEE Southeast Conf.*, 1993.
5. R. C. Garcia and M. N. O. Sadiku, Monte Carlo fixed-radius floating random walk solution for potential problems, *Proc. of IEEE Southeast Conf.*, 1996, pp. 88–91.
6. S. Kakutani, Two dimensional Brownian motion and harmonic functions, *Proc. Imp. Acad.* (Tokyo), vol. 20, 1945, pp. 706–714.
7. J. M. Delanrentis and L. A. Romero, A Monte Carlo method for Poisson's equation, *J. Comp. Phys.*, vol. 90, 123–140, 1990.
8. R. Schlott, A Monte Carlo method for the Dirichlet problem of dielectric wedges, *IEEE Trans. Microwave Theory Tech.*, vol. 36, no. 4, 724–730, April 1988.
9. K. Asano, Electrostatic potential and field in a cylindrical tank containing charged liquid, *Proc. IEE*, vol. 124, no. 12, 1977, pp. 1277–1281.
10. J. H. Pickles, Monte Carlo field calculations, *Proc. of IEEE*, vol. 124, no. 12, December 1977, pp. 1271–1276.
11. M. N. O. Sadiku and K. Gu, Floating random walk algorithm for solving axisymmetric problems, to appear in *Int. J. of Numerical Modeling*.

Problems

5.1. Solve Problem 4.8 (see Figure 4.19) using the floating random walk MCM.

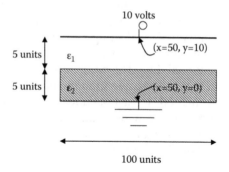

FIGURE 5.7
Figure for Problem 5.2.

5.2. Consider the geometry for a parallel-plate capacitor struc-
ture with two dielectrics, as shown in Figure 5.7. Take $\varepsilon_1 = \varepsilon_o$
and $\varepsilon_2 = 3.9\varepsilon_o$. Calculate the potentials at five different points
$(x, y) = (50, 2)$, $(50, 4)$, $(50, 6)$, $(50, 8)$, and $(50, 9)$. Fringing effects
may be neglected.

5.3. Consider the microstrip geometry shown in Figure 5.8. Take
$\varepsilon_1 = \varepsilon_o$ and $\varepsilon_2 = 3.9\varepsilon_o$. Calculate potential at $(x, y) = (50, 15)$, $(55, 15)$,
$(60, 15)$, $(60, 5)$, and $(50, 5)$.

5.4. Using the floating random walk method, rework Problem 4.12
(Chapter 4).

5.5. Consider the solution region shown Figure 5.9. We want to solve
Poisson's equation

$$\nabla^2 V = -\frac{\rho_v}{\varepsilon} \quad \text{in region R}$$

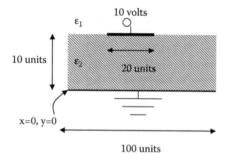

FIGURE 5.8
Figure for Problem 5.3.

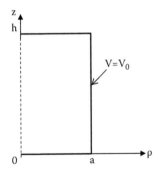

FIGURE 5.9
Figure for Problem 5.5.

subject to Dirichlet boundary conditions

$$V = V_b \text{ on the boundary B}$$

Take $a = h = 1$ m, $\varepsilon_r = 2.5$, $\rho_v = 10^{-5}$ C/m^3. Using the floating random walk method, calculate V at $(0.25, 0.25)$, $(0.75, 0.25)$, $(0.5, 0.5)$, $(0.25, 0.75)$, $(0.75, 0.75)$.

6

The Exodus Method

Kill a man, and you are a murderer. Kill millions of men, and you are a conqueror. Kill everyone, and you are a god.

—Jean Rostand

The Exodus method was first suggested in [1] and was applied to heat problems. It was later developed for electromagnetics [2–3]. Although the method is probabilistic in approach, it is not subject to randomness as are other Monte Carlo methods (MCMs) such as fixed random walk and floating random walk. The Exodus method does not need a random generating routine, and this makes the solution independent of the computing facilities. The method gives the "exact" solution in the sense that the main source of error is in the estimation of the transition probabilities, which are calculated in such a way that randomness is avoided. It has been found that the Exodus method provides solutions that are more accurate than those provided by the fixed or floating MCMs.

In this chapter, we first consider applying the Exodus method to Laplace's equation. We then extend the applicability of the method to Poisson's equation.

6.1 Solution of Laplace's Equation

Applying the Exodus method in finding the solution of a potential problem usually involves the following three steps:

1. We first obtain the random walk probabilities from the finite difference equivalent of the partial differential equation describing the problem.
2. The Exodus method is used along with the random walk probabilities in calculating the transition probabilities.
3. The potential at the point of interest is finally obtained using the transition probabilities and the boundaries conditions.

127

In this section, we apply the Exodus method to Dirichlet problems in rectangular and axisymmetric solution regions.

6.1.1 Rectangular Solution Region

Suppose the Exodus method is to be applied in selecting Laplace's equation

$$\nabla^2 V = 0 \quad \text{in region R} \tag{6.1}$$

subject to the Dirichlet boundary condition

$$V = V_b \quad \text{on boundary B} \tag{6.2}$$

We begin by dividing the rectangular solution region R into a mesh and derive the finite difference equivalent. Assuming a mesh side $\Delta x = \Delta y = \Delta$, the finite difference equivalent of Equation 6.1 is

$$V(x,y) = p_{x+}V(x+\Delta,y) + p_{x-}V(x-\Delta,y) + p_{y+}V(x,y+\Delta) + p_{y-}V(x,y-\Delta) \tag{6.3}$$

where

$$p_{x+} = p_{x-} = p_{y+} = p_{y-} = \frac{1}{4} \tag{6.4}$$

A probabilistic interpretation of Equation 6.3 is that a random-walking particle at an arbitrary point (x, y) in R has probabilities p_{x+}, p_{x-}, p_{y+}, and p_{y-} of moving from (x, y) to the neighboring points $(x + \Delta, y)$, $(x - \Delta, y)$, $(x, y + \Delta)$, and $(x, y - \Delta)$, respectively. Henceforth, p_{x+}, p_{x-}, p_{y+}, and p_{y-} will be referred to as the *random walk probabilities*.

It is also important to consider an extension of the MCM for the solution of problems with discrete homogeneities, that is, homogeneous media separated by interfaces. Consider the interface along the $y = $ constant plane as shown in Figure 6.1. The finite difference equivalent of the boundary condition $D_{1n} = D_{2n}$ at the interface is obtained by applying $\oint D \cdot dS$ on the interface. The result is

$$V_o = p_{x+}V_1 + p_{x-}V_2 + p_{y+}V_3 + p_{y-}V_4 \tag{6.5}$$

where

$$p_{x+} = p_{x-} = \frac{1}{4}, \quad p_{y+} = \frac{\varepsilon_1}{2(\varepsilon_1 + \varepsilon_2)}, \quad p_{y-} = \frac{\varepsilon_2}{2(\varepsilon_1 + \varepsilon_2)} \tag{6.6}$$

An interface along the $x = $ constant plane can be treated in a similar manner.

On the line of symmetry, the condition $\partial V/\partial n = 0$ must be imposed. If the line of symmetry is along the y-axis as in Figure 6.2a,

$$V_o = p_{x+}V_1 + p_{y+}V_3 + p_{y-}V_4 \tag{6.7}$$

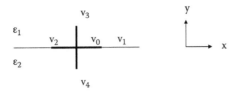

FIGURE 6.1
Interface between media of permittivities ε_1 and ε_2.

where

$$p_{x+} = \frac{1}{2}, \quad p_{y+} = p_{y-} = \frac{1}{4}$$

(6.8)

The line of symmetry along the x-axis is treated similarly.

6.1.2 Axisymmetric Solution Region

For $V = V(\rho, z)$, Equation 6.1 becomes

$$\frac{\partial^2 V}{\partial \rho^2} + \frac{1}{\rho}\frac{\partial V}{\partial \rho} + \frac{\partial^2 V}{\partial z^2} = 0$$

(6.9)

For ease of computation, we assume a square grid so that the step sizes along ρ and z coordinates are equal, that is, $\Delta \rho = \Delta z = \Delta$. The finite difference approximation for $\rho \neq 0$ is

$$V(\rho, z) = p_{\rho+}V(\rho+\Delta, z) + p_{\rho-}V(\rho-\Delta, z) + p_{z+}V(\rho, z+\Delta) + p_{z-}V(\rho, z-\Delta)$$

(6.10)

where

$$p_{z+} = p_{z-} = \frac{1}{4}, \quad p_{\rho+} = \frac{1}{4} + \frac{\Delta}{8\rho}, \quad p_{\rho-} = \frac{1}{4} - \frac{\Delta}{8\rho}$$

(6.11)

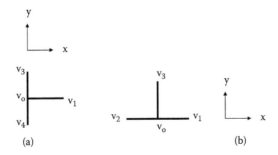

(a)   (b)

FIGURE 6.2
Satisfying symmetry conditions: (a) (b).

Note that the random walk probabilities satisfy

$$p_{z+} + p_{z-} + p_{\rho+} + p_{\rho-} = 1 \tag{6.12}$$

Equations 6.10 and 6.11 do not apply when $\rho = 0$. Because $\partial V/\partial \rho = 0$ at $\rho = 0$, applying L'Hopital's rule yields

$$\lim_{\rho \longrightarrow 0} \frac{1}{\rho} \frac{\partial V}{\partial \rho} = \frac{\partial^2 V}{\partial \rho^2} \tag{6.13}$$

Hence, at $\rho = 0$, Laplace's equation becomes

$$2 \frac{\partial^2 V}{\partial \rho^2} + \frac{\partial^2 V}{\partial z^2} = 0 \tag{6.14}$$

and the finite difference equivalent is

$$V(0,z) = p_{\rho+}V(\Delta,z) + p_{z+}V(0,z+\Delta) + p_{z-}V(0,z-\Delta) \tag{6.15}$$

where

$$p_{z+} = p_{z-} = \frac{1}{6}, \quad p_{\rho+} = \frac{4}{6}, \quad p_{\rho-} = 0 \tag{6.16}$$

For the z = constant interface, the boundary condition $D_{1n} = D_{2n}$ or $\varepsilon_1 \partial V_1/\partial z = \varepsilon_2 \partial V_2/\partial z$ leads to random walk probabilities:

$$p_{z+} = \frac{\varepsilon_1}{2(\varepsilon_1 + \varepsilon_2)}, \quad p_{z-} = \frac{\varepsilon_2}{2(\varepsilon_1 + \varepsilon_2)} \tag{6.17}$$

whereas $p_{\rho+}$ and $p_{\rho-}$ remain as in Equation 6.11. For ρ = constant interface, the boundary condition $(\varepsilon_1/\rho)\partial V_1/\partial \rho = (\varepsilon_2/\rho)\partial V_2/\partial \rho$ leads to

$$p_{\rho+} = \frac{\varepsilon_1}{2(\varepsilon_1 + \varepsilon_2)}\left(1 + \frac{\Delta}{2\rho}\right), \quad p_{\rho-} = \frac{\varepsilon_2}{2(\varepsilon_1 + \varepsilon_2)}\left(1 - \frac{\Delta}{2\rho}\right) \tag{6.18}$$

whereas p_{z+} and p_{z-} remain as in Equation 6.11.

6.1.3   Exodus Method

Suppose that for concreteness we are interested in solving the problem defined in Equations 6.1 and 6.2, and the potential at a specific point (x_o, y_o) is to be determined. We define *the transition probability* p_k as the probability that a random walk starting at the point of interest (x_o, y_o) in R ends at the boundary point (x_k, y_k) with prescribed potential $V_b(k)$, that is,

$$p_k = \text{Prob}(x_o, y_o \longrightarrow x_k, y_k) \tag{6.19}$$

If there are M boundary or fixed nodes (excluding the corner points because a random walk never terminates at those points), the potential at the starting point (x_o, y_o) of the random walks is

$$V(x_o, y_o) = \sum_{k=1}^{M} p_k V_b(k) \qquad (6.20)$$

If m is the number of different boundary potentials, Equation 6.20 can be simplified to

$$V(x_o, y_o) = \sum_{k=1}^{m} p_k V_b(k) \qquad (6.21)$$

where p_k in this case is the probability that a random walk terminates on boundary k. As $V_b(k)$ is specified, our problem is reduced to finding p_k. It is evident from Equation 6.21 that the value of $V(x_o, y_o)$ would be "exact" only if the values of the transition probability p_k are calculated exactly. The values of p_k can be obtained analytically using an expansion technique described in [4]. However, this approach is limited to homogeneous rectangular solution regions. For inhomogeneous or nonrectangular regions, we must resort to a numerical technique. The Exodus method offers a numerical means of finding p_k.

To apply the Exodus method, let $P(i,j)$ be the number of particles at point (i,j) in R. We begin the application of the Exodus method by setting $P(i,j) = 0$ at all nodes (both fixed and free) except at free node (x_o, y_o), where $P(i,j)$ assumes a large number N (say, $N = 10^6$ or more). In other words, we introduce a large number of particles at (x_o, y_o) to start with. By scanning the mesh as is usually done in finite difference analysis, we dispatch the particles at each free node to its neighboring nodes according to the random walk probabilities $p_{x+}, p_{x-}, p_{y+},$ and p_{y-} as illustrated in Figure 6.3. Note that in Figure 6.3b, new $P(i,j) = 0$ at the node, whereas old $P(i,j)$ is shared among the neighboring nodes. At the end of each iteration (i.e., scanning of the free nodes in R as illustrated in Figure 6.3), we record the number of particles that have reached the boundary (i.e., the fixed nodes), where the particles are absorbed. We keep scanning the mesh in a manner similar to the iterative process applied in finite difference solution until a set number of particles (say, 99.99% of N) have reached the boundary. If N_k is the number of particles that have reached boundary k, we calculate

$$p_k = \frac{N_k}{N} \qquad (6.22)$$

Hence, Equation 6.21 can be written as

$$V(x_o, y_o) = \frac{\sum_{k=1}^{m} N_k V_b(k)}{N} \qquad (6.23)$$

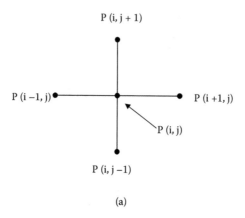

$P(i, j+1)$

$P(i-1, j)$ $P(i+1, j)$

$P(i, j)$

$P(i, j-1)$

(a)

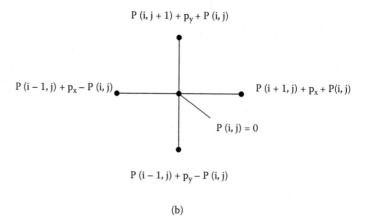

$P(i, j+1) + p_y + P(i, j)$

$P(i-1, j) + p_x - P(i, j)$ $P(i+1, j) + p_x + P(i, j)$

$P(i, j) = 0$

$P(i-1, j) + p_y - P(i, j)$

(b)

FIGURE 6.3
Random walk probabilities: (a) before the particles at (i, j) are dispatched, (b) after the particles at (i, j) are dispatched.

Thus, the problem is reduced to just finding N_k using the Exodus method, given N and $V_b(k)$. We note that if $N \longrightarrow \infty, \Delta \longrightarrow 0$, and all the particles were allowed to reach the boundary points, the values of p_k, and consequently $V(x_o, y_o)$, would be exact. It is easier to approach this exact solution using the Exodus method than using other MCMs and perhaps other numerical techniques such as finite difference and finite element methods. This fact will be illustrated with examples shortly.

For an axisymmetric solution region, we find the transition probability $p_k = \text{Prob}(\rho_o, z_o \longrightarrow \rho_k, z_k)$ in the same way except that at each node we use random walk probabilities p_{z+}, p_{z-}, $p_{\rho+}$, and $p_{\rho-}$ in dispatching the particles.

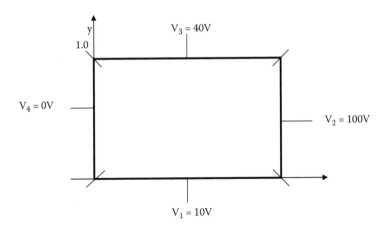

FIGURE 6.4
Figure for Example 6.1.

We shall illustrate the application of the Exodus method by means of two examples. The two examples have an analytic solution so that the accuracy and validity of the Exodus solution can be checked.

Example 6.1
Consider the cross section of an infinitely long trough whose sides are maintained as shown in Figure 6.4. Write MATLAB® code using the Exodus method to calculate the potential at (0.5, 0.5), (0.8, 0.8), (1.0, 0.5), and (0.8, 0.2). Compare your results with exact results [5].

Solution
The MATLAB program is in Figure 6.5, where the step size is selected as $h = \Delta = 0.06$. The code is self-explanatory. The results from the Exodus method are compared with the exact solution as shown in Table 6.1.

Example 6.2
Consider the potential problem shown in Figure 6.6. The potentials at $x = 0$, $x = a$, and $y = 0$ sides are zero, and the potential at $y = b$ side is V_o.

Solution
The analytic solution to this problem using series expansion technique is

$$V = \begin{cases} \displaystyle\sum_{k=1}^{\infty} \sin \beta x \, [a_n \sinh \beta y + b_n \cosh \beta y], & 0 \le y \le c \\[4mm] \displaystyle\sum_{k=1}^{\infty} c_n \sin \beta x \sinh \beta y, & c \le y \le b \end{cases}$$

```
% This program uses the Exodus method to solve Laplace's equation


a=1.5; b=1; h=0.05;

xo=0.8; yo=0.2;

imax=a/h;

jmax=b/h;

io=xo/h;

jo=yo/h;

N = 10^6;  % total no. of particles injected

nmax=500;

P=zeros(imax+1,jmax+1);

P(io+1,jo+1)=N;

sum1=0; v1=10   % bottom side

sum2=0; v2=100 % right side

sum3=0; v3=40   % top side

sum4=0; v4=0    % left side

for n=1:nmax

  % scan the free nodes

  for i=2:imax

  for j=2:jmax

    P(i+1,j)= P(i+1,j) + 0.25* P(i,j);

    P(i-1,j)= P(i-1,j) + 0.25*P (i,j);

    P(i,j+1)= P(i,j+1) + 0.25* P(i,j);
```

FIGURE 6.5
MATLAB code.

```
% This program uses the Exodus method to solve Laplace's equation


a=1.5; b=1; h=0.05;

xo=0.8; yo=0.2;

imax=a/h;

jmax=b/h;

io=xo/h;

jo=yo/h;

N = 10^6;  % total no. of particles injected

nmax=500;

P=zeros(imax+1,jmax+1);

P(io+1,jo+1)=N;

sum1=0; v1=10   % bottom side

sum2=0; v2=100 % right side

sum3=0; v3=40   % top side

sum4=0; v4=0    % left side

for n=1:nmax

  % scan the free nodes

  for i=2:imax

  for j=2:jmax

    P(i+1,j)= P(i+1,j) + 0.25* P(i,j);

    P(i-1,j)= P(i-1,j) + 0.25*P (i,j);

    P(i,j+1)= P(i,j+1) + 0.25* P(i,j);
```

FIGURE 6.5
(Continued).

TABLE 6.1

The Exact Solution and the
Exodus Method

(x, y)	Exact	Exodus
(0.5, 0.5)	22.44	22.44
(0.8, 0.8)	38.55	38.56
(1.0, 0.5)	43.22	43.22
(0.8, 0.2)	20.97	20.91

where

$$\beta = \frac{n\pi}{a}, \quad n = 2k-1,$$

$$a_n = 4V_o[\varepsilon_1 \tanh \beta c - \varepsilon_2 \coth \beta c]/d_n$$

$$b_n = 4V_o(\varepsilon_2 - \varepsilon_1)/d_n$$

$$c_n = 4V_o[\varepsilon_1 \tan \beta c - \varepsilon_2 \coth \beta c + (\varepsilon_2 - \varepsilon_1)\coth \beta c]/d_n$$

$$d_n = n\pi \sinh \beta b[\varepsilon_1 \tanh \beta c - \varepsilon_2 \coth \beta c + (\varepsilon_2 - \varepsilon_1)\coth \beta b]$$

Typically, values

$$V_o = 100, \quad \varepsilon_1 = \varepsilon_o, \quad \varepsilon_2 = 2.25\varepsilon_o, \quad a = 3.0, \quad b = 2.0, \quad c = 1.0$$

are used in all calculations. The potentials are calculated at five typical points using the Exodus method, the fixed random walk MCM, and the analytic solution. The number of particles, N, is taken as 10^7 for the Exodus method, and the step size $\Delta = 0.05$ was used. For the fixed random walk method, $\Delta = 0.05$ and 2000 walks are used. It was noted that 2000 walks are sufficient for the random walk solutions to converge. The results are displayed in Table 6.2. In the table, δ is the error estimate, which is obtained by repeating

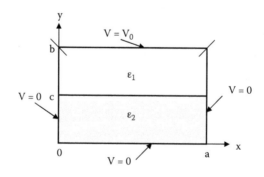

FIGURE 6.6
Potential system for Example 6.2.

TABLE 6.2

Results of Example 6.2

x	y	Exodus Method	Fixed Random Walk $(V \pm \delta)$	Finite Difference	Exact Solution
0.5	1.0	13.41	13.40 ± 1.113	13.16	13.41
1.0	1.0	21.13	20.85 ± 1.612	20.74	21.13
1.5	1.0	23.43	23.58 ± 1.2129	22.99	23.43
1.5	0.5	10.52	10.13 ± 0.8789	10.21	10.51
1.5	1.5	59.36	58.89 ± 2.1382	59.06	59.34

each calculation five times and using the statistical formula provided in Chapter 2, Section 2.5. It should be noted in the table that the results of the Exodus method agree to three significant places with the exact solution. Thus, the Exodus method is more accurate than the random walk technique. It should also be noted that the Exodus method does not require the use of random number routine or error estimate calculation. The Exodus method, therefore, takes less computation time than the random walk method.

6.2 Solution of Poisson's Equation

We now apply the Exodus method to Poisson's equation

$$\nabla^2 V = -g(x,y) = -\frac{\rho_v}{\varepsilon} \quad \text{in region R} \tag{6.24}$$

subject to the Dirichlet boundary condition

$$V = V_b \quad \text{on boundary B} \tag{6.25}$$

where ρ_v (in coulombs/cubic meter) is not a function of z. We seek to solve Poisson's equation for problems involving rectangular and axisymmetric solution regions.

6.2.1 Rectangular Solution Region

We begin by dividing the rectangular solution region R into a mesh and derive the finite difference equivalent of Equation 6.24. Assuming a mesh with $\Delta x = \Delta y = \Delta$, the finite difference equivalent of Equation 6.24 is

$$V(x,y) = p_{x+}V(x+\Delta,y) + p_{x-}V(x-\Delta,y) + p_{y+}V(x,y+\Delta)$$

$$+ p_{y-}V(x,y-\Delta) + \frac{\Delta^2 g}{4} \tag{6.26}$$

where

$$p_{x+} = p_{x-} = p_{y+} = p_{y-} = \frac{1}{4} \tag{6.27}$$

The probabilities p_{x+}, p_{x-}, p_{y+}, and p_{y-} are the *random walk probabilities*. These probabilities can also be obtained for problems with discrete homogeneities (i.e., homogeneous media separated by interfaces) and problems with lines of symmetry. They are derived in Section 6.1 for Laplace's equation and can be derived for Poisson's equation by following a similar procedure.

6.2.2 Axisymmetric Solution Region

For $V = V(\rho, z)$, Equation 6.24 becomes

$$\frac{\partial^2 V}{\partial \rho^2} + \frac{1}{\rho}\frac{\partial V}{\partial \rho} + \frac{\partial^2 V}{\partial z^2} = -g(\rho, z) = -\frac{\rho_v}{\varepsilon} \tag{6.28}$$

where ρ_v is independent of ϕ. To ease computation, we let $\Delta \rho = \Delta z = \Delta$. The finite difference approximation for $\rho \neq 0$ is

$$V(\rho, z) = p_{\rho+} V(\rho + \Delta, z) + p_{\rho-} V(\rho - \Delta, z) + p_{z+} V(\rho, z + \Delta)$$

$$+ p_{z-} V(\rho, z - \Delta) + \frac{\Delta^2 g}{4} \tag{6.29}$$

where

$$p_{z+} = p_{z-} = \frac{1}{4}, \quad p_{\rho+} = \frac{1}{4} + \frac{\Delta}{8\rho}, \quad p_{\rho-} = \frac{1}{4} - \frac{\Delta}{8\rho} \tag{6.30}$$

Equations 6.29 and 6.30 do not apply when $\rho = 0$. Because $\partial V/\partial \rho = 0$ at $\rho = 0$, applying L'Hopital's rule yields

$$\lim_{\rho \longrightarrow 0} \frac{1}{\rho}\frac{\partial V}{\partial \rho} = \frac{\partial^2 V}{\partial \rho^2} \tag{6.31}$$

Hence, at $\rho = 0$, Laplace's equation becomes

$$2\frac{\partial^2 V}{\partial \rho^2} + \frac{\partial^2 V}{\partial z^2} = -g \tag{6.32}$$

and the finite difference equivalent is

$$V(0, z) = p_{\rho+} V(\Delta, z) + p_{z+} V(0, z + \Delta) + p_{z-} V(0, z - \Delta) + \frac{\Delta^2 g}{6} \tag{6.33}$$

where

$$p_{z+} = p_{z-} = \frac{1}{6}, \quad p_{\rho+} = \frac{4}{6}, \quad p_{\rho-} = 0 \tag{6.34}$$

The random walk probabilities p_{z+}, p_{z-}, $p_{\rho+}$, and $p_{\rho-}$ are the probabilities that a random-walking particle at an arbitrary point (ρ,z) in R will move to the neighboring points $(\rho, z + \Delta)$, $(\rho, z - \Delta)$, $(\rho + \Delta, z)$, and $(\rho - \Delta, z)$, respectively. Again, these probabilities are derived for Laplace's equations in Section 6.1 for problems with discrete homogeneities and lines of symmetry, and can be derived in a similar manner for Poisson's equation.

6.2.3 Transition and Transient Probabilities

To compute the solution of the problem defined in Equation 6.25, for example, at a specific point (x_o, y_o), we need the *transition probability* p_k and the *transient probability* q_ℓ. The transition probability p_k is already defined as the probability that a random walk starting at the point of interest (x_o, y_o) in R ends at a boundary point (x_k, y_k), where potential $V_b(k)$ is prescribed, that is,

$$p_k = \text{Prob}(x_o, y_o \longrightarrow x_k, y_k) \tag{6.35}$$

The transient probability q_ℓ is the probability that a random walk starting at point (x_o, y_o) passes through point (x_ℓ, y_ℓ) on the way to the boundary, that is,

$$q_\ell = \text{Prob}(x_o, y_o \xrightarrow{x_\ell, y_\ell} \text{boundary B}) \tag{6.36}$$

If there are m boundary (or fixed) nodes (excluding the corner points because a random walk never terminates at those points) and M_f free nodes in the mesh, the potential at the starting point (x_o, y_o) of the random walks is

$$V(x_o, y_o) = \sum_{k=1}^{m} p_k V_b(k) + \sum_{\ell=1}^{M_f} q_\ell G_\ell \tag{6.37}$$

where

$$G_\ell = \frac{\Delta^2}{4} g(x_\ell, y_\ell) \tag{6.38}$$

If M_b is the number of different boundary potentials, the first term in the right-hand side of Equation 6.37 can be simplified so that

$$V(x_o, y_o) = \sum_{k=1}^{M_b} p_k V_b(k) + \sum_{\ell=1}^{M_f} q_\ell G_\ell \tag{6.39}$$

where p_k in this case is the probability that a random walk terminates on boundary k. As $V_b(k)$ is specified and the source term G_ℓ is known, our problem is reduced to finding the probabilities p_k and q_ℓ. We note from Equation 6.39 that the value of $V(x_o, y_o)$ would be exact if the transition probabilities p_k and the transient probabilities q_ℓ were known exactly. These probabilities can be obtained in one of two ways: either analytically or numerically. The analytic approach involves using an expansion technique described in [4]. However, this approach is limited to homogeneous rectangular solution regions. For inhomogeneous or nonrectangular regions, we must resort to some numerical simulation. The Exodus method offers a numerical means of finding p_k and q_ℓ. The fixed random walk can also be used to compute the transient and transition probabilities.

6.2.4 Exodus Method

Suppose it is desired to find V at point (x_o, y_o). To apply the Exodus method, let $P(i, j)$ be the number of particles at point (i, j) in R, and $Q(i, j)$ the number of particles passing through the same point. We begin the application of the Exodus method by setting $P(i, j) = 0 = Q(i, j)$ at all nodes (both fixed and free) except at the free node (x_o, y_o), where $P(i, j)$ and $Q(i, j)$ are set equal to a large number N_p (say, $N_p = 10^6$ or more). In other words, we inject a large number of particles at (x_o, y_o) to start with. By scanning the mesh iteratively as is usually done in finite difference analysis, we dispatch the particles at each free node to its neighboring nodes according to the random walk probabilities $p_{x+}, p_{x-}, p_{y+},$ and $p_{y-},$ as illustrated in Figure 6.3. Note that in Figure 6.3b, new $P(i, j) = 0$ at that node, whereas old $P(i, j)$ is shared among the neighboring nodes. As shown in Figure 6.7, the value of $Q(i, j)$ does not change at that node, whereas Q at the neighboring nodes is increased by the old $P(i, j)$ that is shared by those nodes. Although $P(i, j)$ keeps records of the number of particles at point (i, j) during each iteration, $Q(i, j)$ tallies the number of particles passing through that point.

At the end of each iteration (i.e., scanning of the free nodes in R as illustrated in Figures 6.3 and 6.7), we record the number of particles that have reached the boundary (i.e., the fixed nodes) where the particles are absorbed. We keep scanning the mesh in a manner similar to the iterative process applied in finite difference solution until a set number of particles (say, 99.99% of N_p) have reached the boundary. If N_k is the number of particles that reached boundary k, we calculate

$$p_k = \frac{N_k}{N_p} \tag{6.40}$$

Also, at each free node, we calculate

$$q_\ell = \frac{Q_\ell}{N_p} \tag{6.41}$$

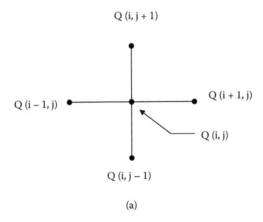

Q (i, j + 1)

Q (i − 1, j)

Q (i + 1, j)

Q (i, j)

Q (i, j − 1)

(a)

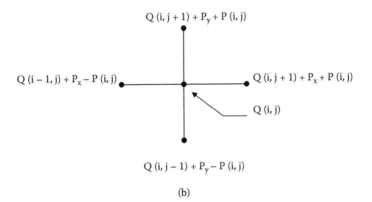

Q (i, j + 1) + P$_y$ + P (i, j)

Q (i − 1, j) + P$_x$ − P (i, j)

Q (i, j + 1) + P$_x$ + P (i, j)

Q (i, j)

Q (i, j − 1) + P$_y$ − P (i, j)

(b)

FIGURE 6.7
Number of particles passing through node (i, j) and its neighboring nodes: (a) before the particles at the node are dispatched, (b) after the particles at the node are dispatched.

where $Q_\ell = Q(i,j)$ is now the total number of particles that have passed through that node on their way to the boundary. Hence, Equation 6.39 can be written as

$$V(x_o, y_o) = \frac{\sum_{k=1}^{M_b} N_k V_b(k)}{N_p} + \frac{\sum_{\ell=1}^{M_f} Q_\ell G_\ell}{N_p} \tag{6.42}$$

Thus, the problem is reduced to just finding N_k and Q_ℓ using the Exodus method, given N_p, $V_b(k)$, and G_ℓ. If $N_p \longrightarrow \infty$, $\Delta \longrightarrow 0$, and all the particles were allowed to reach the boundary points, the values of p_k and q_ℓ and consequently $V(x_o, y_o)$ would be exact. It is easier to approach this exact solution using the Exodus method than using other MCMs and perhaps other numerical techniques such as finite difference and finite element methods.

6.2.5 Fixed Random Walk

The fixed random walk Monte Carlo technique can be used as an alternative numerical approach for calculating the transient and transition probabilities. A detailed description of the random walk MCM is given in Chapter 4. To find the transition probability p_k and transient probability q_ℓ in Equation 6.39, the technique essentially involves taking the following steps:

1. Start each random walk at (x_o, y_o).
2. Generate a random number U that is uniformly distributed between 0 and 1, and move to a new point according to:

$$(x,y) \longrightarrow (x+\Delta, y) \quad \text{if } 0 < U < 0.25$$

$$(x,y) \longrightarrow (x-\Delta, y) \quad \text{if } 0.25 < U < 0.5$$

$$(x,y) \longrightarrow (x, y+\Delta) \quad \text{if } 0.5 < U < 0.75$$

$$(x,y) \longrightarrow (x, y-\Delta) \quad \text{if } 0.75 < U < 1.0$$

    Increase the value of N_ℓ by one, the number of random walks passing through that point.
3. If the new point is not on the boundary B, repeat step 2. If the new point is on B, terminate the random walk and record the prescribed potential V_b at the boundary point for the i-th random walk. Increase by one the value of N_k, the number of random walks reaching the boundary.
4. Repeat steps 1–3 for N_w random walks, and finally calculate

$$p_k = \frac{N_k}{N_w} \tag{6.43}$$

$$q_\ell = \frac{N_\ell}{N_w} \tag{6.44}$$

The rate of convergence of the solution varies as $\sqrt{N}$, so that many random walks are necessary to ensure accurate results. The values of p_k and q_ℓ in Equations 6.43 and 6.44 are substituted in Equation 6.39 to obtain

$$V(x_o, y_o) = \frac{\sum\limits_{k=1}^{M_b} N_k V_b(k)}{N_w} + \frac{\sum\limits_{\ell=1}^{M_f} N_\ell G_\ell}{N_w} \tag{6.45}$$

The fixed random walk is capable of producing results of the same degree of accuracy as the Exodus method but requires large value of N_w and, consequently, more computation time.

For an axisymmetric solution region, the transition probability $p_k = \text{Prob}$ $(\rho, z \longrightarrow \rho_k, z_k)$ and the transient probability $q_\ell = \text{Prob}(\rho, z \xrightarrow{\rho_\ell, z_\ell}$ boundary B) are found in the same way except that at each node we use the random walk probabilities $p_{z+}, p_{z-}, p_{\rho+}$, and $p_{\rho-}$ in dispatching the particles or taking the random walks as the case may be.

The following two examples are used to illustrate the solution of Poisson's equation by the Exodus method. The solution region of the first example is rectangular, whereas that of the second example is axisymmetric. The two examples have exact analytic solutions so that accuracy and validity of the numerical technique can be checked.

Example 6.3

Consider Poisson's equation

$$\nabla^2 V = -g(x,y) = -36\pi x(y-1), \quad 0 \le x, y \le 1 \tag{6.46}$$

The analytic solution to this problem using series expansion technique is given in [5], where the finite difference solution is also available.

Solution

The potentials were calculated at five typical points using Equation 6.42 for the Exodus method, using Equation 6.45 for the fixed random walk MCM, and the analytic solution. The number of particles, N_p, was taken as 10^6 for the Exodus method, and the step size $\Delta = 0.05$ was used. The iteration was terminated when 0.001% of N_p particles were yet to reach the boundary. This took 600 iterations. For the fixed random walk method, $\Delta = 0.05$ and $N_w = 10{,}000$ walks were used. It was noted that 10,000 walks were sufficient for the random walk solution to converge. The results are displayed in Table 6.3. In this table, δ is the error estimate, which is obtained by repeating each calculation five times and using statistical formulas provided in Chapter 2. The results in Table 6.3 show that the Exodus method is as accurate as the finite difference method. Also, the results from the Exodus method are not

TABLE 6.3

Results of Example 6.3

x	y	Exodus Method	Fixed Random Walk ($V \pm \delta$)	Finite Difference	Exact Solution
0.25	0.25	−3.431	−3.402 ± 0.106	−3.424	−3.425
0.75	0.25	2.880	2.908 ± 0.1073	2.909	2.902
0.5	0.5	2.9078	2.993 ± 0.1127	2.921	2.913
0.25	0.75	4.274	4.242 ± 0.2394	4.277	4.277
0.75	0.75	11.114	11.146 ± 0.1369	11.23	11.131

subject to randomness and consequently do not require calculating the error estimate as do the results from the fixed random walks.

Example 6.4

Consider an earthed metal cylindrical tank partly filled with a charged liquid, such as hydrocarbons. The problem involves solving

$$\nabla^2 V = 0 \quad \text{for gas space}$$

$$\nabla^2 V = -\frac{\rho_v}{\varepsilon} \quad \text{for liquid space} \tag{6.47}$$

Solution

The problem is two dimensional with an axisymmetric solution domain. Therefore, it is only necessary to investigate the solution region in Figure 6.8 and impose the condition that the z-axis is a flux line, that is, $\partial V/\partial \rho = 0$.

By using the series expansion technique, the exact solution is obtained as in Example 4.3. Typically,

$$a = b = c = 1.0 \text{ m}$$

$$\varepsilon_r = 2.0 \text{ (hydrocarbons)} \tag{6.48}$$

$$\rho_v = 10^{-5} \text{ C/m}^3$$

were used in all calculations.

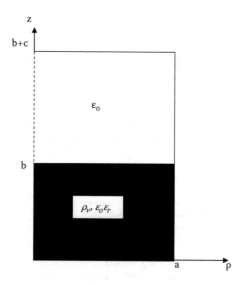

FIGURE 6.8
Figure for Example 6.4.

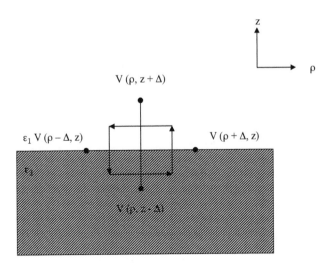

FIGURE 6.9
A small path at the interface separating two media.

To solve this problem using the Exodus method, it should be noted that Equations 6.30 and 6.34 do not apply along the line of symmetry for $z \leq b$, at the corner point $(\rho, z) = (0, b)$, and at the interface $z = b$. By applying Gauss's law

$$\int_v \rho_v dv = -\oint_S \varepsilon \frac{\partial V}{\partial n} dS \qquad (6.49)$$

At the interface, as shown in Figure 6.9, we obtain

$$\left(-\rho_{v1}\frac{\Delta^2}{2} - \rho_{v2}\frac{\Delta^2}{2}\right)\rho\Delta\phi = \left(\varepsilon_1\left[\frac{V(\rho, z+\Delta) - V(\rho, z)}{\Delta}\right]\Delta\right.$$

$$+\varepsilon_1\left[\frac{V(\rho-\Delta, z) - V(\rho, z)}{\Delta}\right]\frac{\Delta}{2} + \varepsilon_2\left[\frac{V(\rho-\Delta, z) - V(\rho, z)}{\Delta}\right]$$

$$+\varepsilon_2\left[\frac{V(\rho, z-\Delta) - V(\rho, z)}{\Delta}\right]\Delta + \varepsilon_2\left[\frac{V(\rho+\Delta, z) - V(\rho, z)}{\Delta}\right]$$

$$\left. +\varepsilon_1\left[\frac{V(\rho+\Delta, z) - V(\rho, z)}{\Delta}\right]\frac{\Delta}{2}\right)\rho\Delta\phi \qquad (6.50)$$

In our particular case, $\rho_{v1} = 0$, $\rho_{v2} = \rho_v$ so that Equation 6.50 becomes

$$V(\rho, z) = p_{\rho+}V(\rho+\Delta, z) + p_{\rho-}V(\rho-\Delta, z) + p_{z+}V(\rho, z+\Delta$$

$$+ p_{z-}V(\rho, z-\Delta) + \frac{\Delta^2\rho_v}{4(\varepsilon_1 + \varepsilon_2)} \qquad (6.51)$$

TABLE 6.4

Random Walk Probabilities and Source Term for Example 6.4

Location	$P_{\rho+}$	$P_{\rho-}$	P_{z+}	P_{z-}	G_ℓ
$z > b, \rho \neq 0$	$\dfrac{1}{4} + \dfrac{\Delta}{8\rho}$	$\dfrac{1}{4} - \dfrac{\Delta}{8\rho}$	$1/4$	$1/4$	0
$z < b, \rho \neq 0$	$\dfrac{1}{4} + \dfrac{\Delta}{8\rho}$	$\dfrac{1}{4} - \dfrac{\Delta}{8\rho}$	$1/4$	$1/4$	$\dfrac{\rho_v \Delta^2}{4\varepsilon_2}$
$z = b, \rho \neq 0$	$\dfrac{1}{4} + \dfrac{\Delta}{8\rho}$	$\dfrac{1}{4} - \dfrac{\Delta}{8\rho}$	$\dfrac{\varepsilon_1}{2(\varepsilon_1 + \varepsilon_2)}$	$\dfrac{\varepsilon_2}{2(\varepsilon_1 + \varepsilon_2)}$	$\dfrac{\rho_v \Delta^2}{4(\varepsilon_1 + \varepsilon_2)}$
$z > b, \rho = 0$	$2/3$	0	$1/6$	$1/6$	0
$z < b, \rho = 0$	$2/3$	0	$1/6$	$1/6$	$\dfrac{\rho_v \Delta^2}{12\varepsilon_2}$
$z = b, \rho = 0$	$2/3$	0	$\dfrac{\varepsilon_1}{3(\varepsilon_1 + \varepsilon_2)}$	$\dfrac{\varepsilon_2}{3(\varepsilon_1 + \varepsilon_2)}$	$\dfrac{\rho_v \Delta^2}{12(\varepsilon_1 + \varepsilon_2)}$

where

$$P_{z+} = \frac{\varepsilon_1}{2(\varepsilon_1 + \varepsilon_2)}, \quad P_{z-} = \frac{\varepsilon_2}{2(\varepsilon_1 + \varepsilon_2)}, \quad P_{\rho+} = P_{\rho-} = \frac{1}{4} \tag{6.52}$$

and $\varepsilon_1 = \varepsilon_o, \varepsilon_2 = 2\varepsilon_o$. By following the same procedure, the random walk probabilities can be derived for the line of symmetry for $z \leq b$ and at the corner point $(\rho, z) = (0, b)$. These are conveniently listed in Table 6.4.

For the Exodus method, $\Delta = 0.05$ and $N_p = 10^6$ were used. The iteration was terminated when 0.001% of N_p particles were yet to reach the boundary. For the fixed random walk method, $\Delta = 0.05$ and $N_w = 10,000$ walks were used. Table 6.5 presents the results of the exact solution, Exodus method, and fixed random walk for five typical points.

TABLE 6.5

Results of Example 6.4 (V is in kilovolts)

ρ	z	Exodus Method (V)	Fixed Random walk (V $\pm \delta$)	Exact Solution (V)
0.25	1.0	71.366	71.367 ± 0.5464	71.12
0.5	1.0	58.185	58.114 ± 0.5567	58.159
0.75	1.0	34.957	34.764 ± 0.5858	36.02
0.5	0.75	16.980	16.772 ± 0.4533	16.945
0.5	0.5	63.837	63.682 ± 0.4485	63.88

6.3 Summary

The Exodus method provides a relatively straightforward means of solving potential problems. The method provides a more accurate solution in less time compared with the fixed random walk.

It is interesting to note that the accuracy of the Exodus method does not really depend on the number of particles N_p. The accuracy depends on the step size Δ and the number of iterations or the tolerance, say, 0.001% of N_p are yet to reach the boundary before the iteration is terminated. However, a large value of N_p reduces the truncation error in the computation.

Although the Exodus method is probabilistic in its approach, it is not subject to randomness as other Monte Carlo techniques because it does not involve the use of a pseudorandom generation subroutine. However, as with other MCMs, it still requires calculating the potential at one point at a time. Concurrent programming techniques have been recently been used to alleviate this problem [6]. Using Markov chains is another way to overcome the problem, as we will see Chapter 8.

References

1. A. F. Emery and W. W. Carson, A modification to the Monte Carlo method—the Exodus method, *J. Heat Trans. ASME*, Series C, vol. 90, 328–332, 1968.
2. M. N. O. Sadiku and D. Hunt, Solution of Dirichlet problems by the Exodus method, *IEEE Trans. Microwave Theory Tech.*, vol. 40, no. 1, 89–95, January 1992.
3. M. N. O. Sadiku. S. O. Ajose, and Z. Fu, Applying the Exodus method to solve Poisson's equation, *IEEE Trans. Microwave Theory Tech.*, vol. 42, no. 4, 661–666, April 1994.
4. W. H. McCrea and F. J. W. Whipple, Random paths in two and three dimensions, *Proc. Roy. Soc. Edinburgh*, vol. 60, 1940, pp. 281–298.
5. M. N. O. Sadiku, *Numerical Techniques in Electromagnetics*, 2nd ed., Boca Raton, FL: CRC Press, 2001, pp. 35–38, 86, 87.
6. V. F. Fusco, Concurrent processing the key to rapid probabilistic potential method, *Int. J. Numerical Modeling Electronic, Networks, Devices, and Field*, vol. 3, 1–9, 1990.

Problems

6.1. Given the one-dimensional differential equation

$$\frac{d^2\Phi}{dx^2} = 0, \quad 0 \le x \le 1$$

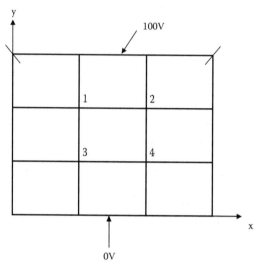

FIGURE 6.10
Figure for Problem 6.2.

subject to $\Phi(0) = 0$, $\Phi(1) = 10$, use the Exodus method to find $\Phi(0.25)$ by injecting 256 particles at $x = 0.26$. You can solve this problem by hand calculation.

6.2. Use the Exodus method to find the potential at node 4 in Figure 6.10. Inject 256 particles at node 4, and scan nodes in the order 1, 2, 3, 4. You can solve this problem by hand calculation.

6.3. Write a program that will apply the Exodus method to determine the potential at point $(0.2, 0.4)$ in the system shown in Figure 6.11.

6.4. Consider the potential problem shown in Figure 6.12. The potentials at $x = 0$, $x = w$, and $y = 0$ sides are zero, and the potential at

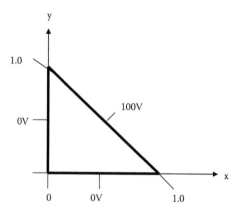

FIGURE 6.11
Figure for Problem 6.3.

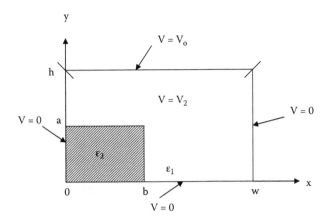

FIGURE 6.12
Figure for Problem 6.4.

$y = h$ side is V_o. Solve the problem using the Exodus method with $V_o = 100$, $\varepsilon_1 = \varepsilon_o$, $\varepsilon_2 = 3\varepsilon_o$, $a = b = 0.5$, $h = w = 1.0$. At the corner point $(x,y) = (a,b)$, use

$$p_{x+} = p_{x-} = \frac{\varepsilon_1}{3\varepsilon_1 + \varepsilon_2}, \quad p_{y+} = p_{y-} = \frac{(\varepsilon_1 + \varepsilon_2)}{2(3\varepsilon_1 + \varepsilon_2)}$$

Calculate the potential at points $(x, y) = (0.25, 0.5)$, $(0.5, 0.5)$, $(0.75, 0.5)$, $(0.5, 0.75)$, and $(0.5, 0.25)$.

6.5. Figure 6.13 depicts a prototype of an electrostatic particle-focusing system employed in a recoil-mass time-of-flight spectrometer.

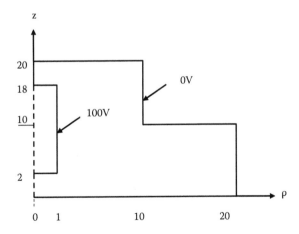

FIGURE 6.13
Figure for Problem 6.5.

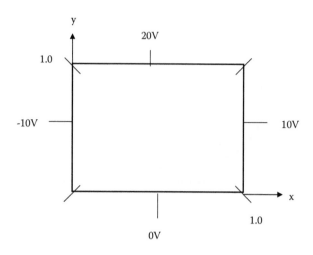

FIGURE 6.14
Figure for Problem 6.6.

Use the Exodus method to find the potential at $(\rho,z) = (5,18)$, $(5,10)$, $(5, 2)$, $(10, 2)$, and $(15, 2)$.

6.6. In the solution region of Figure 6.14, $\rho_v = x(y-1)$ nC/m³. Find the potential at the center of the region using the Exodus method.

7

Neumann Problems

There are two things to aim at in life: first, to get what you want; and after that to enjoy it. Only the wisest of mankind achieve the second.

—Logan P. Smith

As alluded to earlier in this book, there are different kinds of Monte Carlo Methods. Popular ones include the fixed random walk, the floating random walk, and the Exodus method. Until now, in the fixed random walk, the steps of the walks are constrained to lie parallel to the coordinate axes and the meshes are square. The problem with this and other classical Monte Carlo techniques is that these classical methods cannot handle problems involving mixed or Neumann boundary conditions. Handling Neumann problems requires the introduction of another Monte Carlo technique called the *equilateral triangular random walk*. Although this method can also handle Dirichlet problems [1], it is its ability to handle Neumann problems [2] that is the focus of this chapter.

7.1 Governing Equations

The solution region is divided into many equilateral triangular meshes, as shown in Figure 7.1. Each point in the interior solution region has six neighboring nodes, which are equidistant from the central point. When these points are located in different regions, there are different transient probabilities to direct the random walk issued from the central point.

For a concrete example, consider Poisson's equation with mixed boundary conditions:

$$\nabla^2 V(X) = -g(X), \quad X \in \Omega$$

$$V(X) = f_1(X), \quad X \in \partial\Omega_1$$

$$\frac{\partial V(X)}{\partial n} = f_2(X), \quad X \in \partial\Omega_2$$

$$\partial\Omega_1 \cup \partial\Omega_2 = \partial\Omega, \quad \partial\Omega_1 \cap \partial\Omega_2 = 0, \tag{7.1}$$

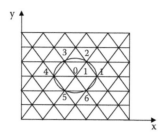

FIGURE 7.1
Triangular meshes.

where $\frac{\partial}{\partial n}$ is the directional derivative along n. When $g(X) = 0$, Equation 7.1 reduces to Laplace's equation.

At point 0 in Figure 7.1, the transient probability equation is [1]

$$V_0 = \frac{1}{6}[V_1 + V_2 + V_3 + V_4 + V_5 + V_6] \tag{7.2}$$

The probabilistic meaning of Equation 7.2 can be explained as follows: The probability of random walk from the central point to any one of the neighboring points is 1/7. A solution for Poisson's equation on Dirichlet boundary condition is obtained for two dimensions as

$$V(X_0) = \frac{1}{N}\sum_{i=1}^{N} V_p(i) + \frac{1}{4N}\sum_{j=1}^{N}\sum_{i=1}^{n_j} g(x_{ij}, y_{ij})d^2 \tag{7.3}$$

where d is the size of the equilateral triangular mesh, n_j is the number of steps in the j-th random walk, and N is the total number of random walks. For Laplace's equation, the second term in Equation 7.3 vanishes.

On the dielectric boundary between two media as shown in Figure 7.2, we apply the boundary condition $D_{1n} = D_{2n}$. This is based on Gauss's law:

$$0 = \oint_L \varepsilon \frac{\partial V}{\partial n} dl \tag{7.4}$$

When we apply this across the interface in Figure 7.2, we obtain

$$0 = \left[\varepsilon_2 \frac{V_1 - V_0}{d}\frac{2\pi d}{12} + \varepsilon_1 \frac{V_1 - V_0}{d}\frac{2\pi d}{12}\right] + \left[\varepsilon_1 \frac{V_2 - V_0}{d}\frac{2\pi d}{6} + \varepsilon_1 \frac{V_3 - V_0}{d}\frac{2\pi d}{6}\right]$$

$$+ \left[\varepsilon_1 \frac{V_4 - V_0}{d}\frac{2\pi d}{12} + \varepsilon_2 \frac{V_4 - V_0}{d}\frac{2\pi d}{12}\right] + \left[\varepsilon_2 \frac{V_5 - V_0}{d}\frac{2\pi d}{6} + \varepsilon_2 \frac{V_6 - V_0}{d}\frac{2\pi d}{6}\right] \tag{7.5}$$

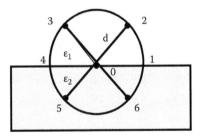

FIGURE 7.2
Interface between two media.

Rearranging the terms gives

$$V_0 = \frac{1}{6}(V_1 + V_4) + \frac{\varepsilon_1}{3(\varepsilon_1 + \varepsilon_2)}(V_2 + V_3) + \frac{\varepsilon_2}{3(\varepsilon_1 + \varepsilon_2)}(V_5 + V_6) \qquad (7.6)$$

Note that the discrete inhomogeneity does not affect the transient probabilities at points 1 and 4 on the boundary but does affect the transient probabilities at points 2, 3, and 4, 5 in proportion to their corresponding permittivities. Also, note that when $\varepsilon_1 = \varepsilon_2$, Equation 7.6 reduces to Equation 7.2.

7.2 Triangular Mesh Method

7.2.1 One Corner on the Boundary

For the triangular mesh in Figure 7.3, a linear equation $V = Ax + By + C$ is constructed. When one or two corners are on the boundary, there are different equations to determine the coefficients A, B, and C. We suppose that

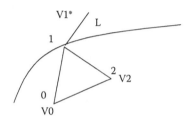

FIGURE 7.3
One corner of the triangle on the boundary.

corner 1 of the triangle lies on the boundary and $V_1^* = \frac{\partial V}{\partial l}$, the directional derivative along $\mathbf{L_1} = (m_1, l_1)$, touches the triangle at point 1. The values at points 0 and 2 are V_0 and V_2, respectively. Thus

$$V_0 = Ax_0 + By_0 + C$$

$$V_1^* = Am_1 + Bl_1$$

$$V_2 = Ax_2 + By_2 + C \tag{7.7}$$

If $\mathbf{L_1}$ is not parallel to side 02 of the triangle, then A, B, and C can be determined. Also, the value of the potential at point 1 is

$$V_1 = Ax_1 + By_1 + C \tag{7.8}$$

7.2.2   Two Corners on the Boundary

If corners 1 and 2 are on the boundary (Figure 7.4), the corresponding equations are

$$V_0 = Ax_0 + By_0 + C$$

$$V_1^* = Am_1 + Bl_1$$

$$V_2^* = Am_2 + Bl_2 \tag{7.9}$$

and also

$$V_1 = Ax_1 + By_1 + C$$

$$V_2 = Ax_2 + By_2 + C \tag{7.10}$$

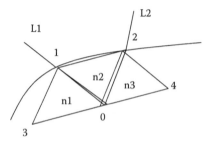

FIGURE 7.4
Two corners of the triangle on the boundary.

where V_1^* is the directional derivative along $\mathbf{L}_1 = (m_1, l_1)$, and V_2^* is the directional derivative along $\mathbf{L}_2 = (m_2, l_2)$. If $\mathbf{L}_1$ and $\mathbf{L}_2$ are not parallel, the coefficients A, B, and C can be determined. An iterative method is used to calculate the potentials.

7.3 Computing Procedure

On the basis of the previous discussion, we can solve Laplace's or Poisson's equation by performing the following steps:

1. The domain is divided into equilateral triangular meshes. Each boundary point is assigned its boundary value or directional derivative value. Each free or exterior point (such as 3, 0, 4 in Figure 7.4) is assigned some initial value.

2. According to the linear function $V = Ax + By + C$, the boundary value of the Neumann boundary is calculated. When directional vectors $\mathbf{L}_1 \| \mathbf{L}_2$, the linear equation on mesh n_2 cannot be determined directly from points 0, 1, 2, because Equation 7.7 has no solution. The values on points 1 and 2 (in Figure 7.4) are obtained from meshes n_1 and n_3.

3. Using the random walk method (Equation 7.3), every interior point or subboundary point value (e.g., point 4 in Figure 7.1) is determined.

   a. If all points are located in the same media, the six neighboring points have the same probability (1/6) of absorbing the random-walking particle. We generate a random number U between 0 and 1. If a random walk is at X_i, then the next step is at X_{i+1}, where

$$x_{i+1} = x_i + d, \; y_{i+1} = y_i, \quad 0 \le U \le 1/6$$

$$x_{i+1} = x_i + \frac{1}{2}d, \; y_{i+1} = y_i + \frac{\sqrt{3}}{2}d, \quad 1/6 \le U \le 1/3$$

$$x_{i+1} = x_i - \frac{1}{2}d, \; y_{i+1} = y_i + \frac{\sqrt{3}}{2}d, \quad 1/3 \le U \le 1/2$$

$$x_{i+1} = x_i - d, \; y_{i+1} = y_i, \quad 1/2 \le U \le 2/3$$

$$x_{i+1} = x_i - \frac{1}{2}d, \; y_{i+1} = y_i - \frac{\sqrt{3}}{2}d, \quad 2/3 \le U \le 5/6$$

$$x_{i+1} = x_i + \frac{1}{2}d, \; y_{i+1} = y_i - \frac{\sqrt{3}}{2}d, \quad 5/6 \le U \le 1 \qquad (7.11)$$

b. If the central point is set on the dielectric boundary (see
Figure 7.2), the random walk is according to Equation 7.6; that is,
the random walk probability along the boundary is 1/6; points
2 and 3 and points 5 and 6 have probabilities of absorbing a
random-walking particle as $\frac{\varepsilon_1}{3(\varepsilon_1+\varepsilon_2)}$ and $\frac{\varepsilon_2}{3(\varepsilon_1+\varepsilon_2)}$, respectively. The
boundary point contribution is

$$\Delta V = \frac{1}{8}[g_1(x_i,y_i)+g_2(x_i,y_i)]d^2 \quad \text{for 2D}$$

$$\Delta V = \frac{1}{12}[g_1(x_i,y_i,z_i)+g_2(x_i,y_i,z_i)]d^2 \quad \text{for 3D} \tag{7.12}$$

Comparing the computed values in two consecutive times and correcting
the boundary value (of the Neumann boundary condition) according to linear equation, the computational procedure is repeated until the difference
between the two computations reaches a prescribed tolerance.

The following two examples illustrate the application of the method presented in this chapter. The first example involves a Laplace's equation with
Neumann boundary condition, and the second one deals with Poisson's
equation with Neumann condition. The first example has an analytic solution so that the accuracy and validity of the triangular mesh random walk
method can be checked. In the second example, the solution is compared
with that of the finite difference method. The two examples were calculated
on an IBM PC. The computer programs for the two examples (in Fortran) can
be found in [3].

Example 7.1

Consider the following Laplace's equation with Neumann boundary condition:

$$\nabla^2 V = 0, \quad 0 < x < \pi, 0 < y < \pi$$

$$V_x(0,y) = 0, \qquad V_x(\pi,y) = 0$$

$$V_y(x,0) = \cos x, \qquad V_y(x,\pi) = 0 \tag{7.13}$$

Solution

The analytic solution of this problem is

$$V(x,y) = \left[\sinh y - \frac{\cosh y}{\tanh \pi} \right] \cos x \tag{7.14}$$

The problem was solved by using the triangular mesh random walk method.
In the example, the triangular mesh side length d is 0.314 and the number of

TABLE 7.1

Solution Results for Example 7.1

x	y	Triangular Mesh Random Walk	Exact Solution
0.471	0.272	−0.6987	−0.6838
1.150	0.544	−0.2192	−0.2300
1.718	1.360	0.03747	0.03800
2.300	1.904	0.10920	0.10820
2.980	2.448	0.09812	0.10500

trials N is 5000. In general, when mesh size is small, a small value may be chosen for the number of trials N. Table 7.1 compares the triangular mesh random walk with the exact analytic solution in Equation 7.14.

Example 7.2

Consider the following Poisson's equation with Neumann boundary conditions:

$$\nabla^2 V = -100(x - 0.5)(y - 0.5), \quad 0 < x < 1, 0 < y < 1$$

$$V_x(0, y) = 0, \qquad V_x(1, y) = 0$$

$$V_y(x, 0) = 0, \qquad V_y(x, 1) = 0 \tag{7.15}$$

Solution

We use the triangular mesh random walk method to determine the potential distribution at several points. In this example, $N = 15000$ and the triangular mesh size $d = 0.1$. The same problem is solved by finite difference method with $\Delta = 0.1$. The results are displayed in Table 7.2.

TABLE 7.2

Solution Results for Example 7.2

x	y	Triangular Mesh Random Walk	Finite Difference
0.2	0.174	0.2961	0.2879
0.2	0.697	−0.088	−0.086
0.8	0.174	−0.139	−0.137
0.8	0.697	0.2466	0.2274
0.4	0.348	0.1332	0.1253

7.4 Summary

This chapter presents a triangular mesh fixed random walk method for solving potential problems in general, especially those with Neumann boundary conditions. For each triangular mesh, a linear function is constructed and expressed by a directional derivative and a potential value on the boundary and exterior points. The two examples considered in this chapter show that the triangular random walk method provides solutions that always converge to the exact solution.

References

1. K. Gu and M. N. O. Sadiku, A triangular mesh random walk for Dirichlet problems, *J. Franklin Institute*, vol. 332B, no. 5, 569–578, 1995.
2. M. N. O. Sadiku and K. Gu, A new Monte Carlo method for Neumann problems, *Proc. IEEE Southeastcon*, 1996, pp. 92–97.
3. K. Gu, Monte Carlo Solutions for Potential and Waveguide Problems, Doctoral dissertation, Temple University, August 1996, pp. 101–111.

Problems

7.1. Use the Monte Carlo method to solve

$$U_{xx} + U_{yy} = 0, \quad 0 < x < 1, \quad 0 < y < 1$$

$$U_x(0,y) = 0, \quad U_x(1,y) = 0, \quad U(x,0) = x^2, \quad U_y(x,1) = 0$$

7.2. Using the Monte Carlo method, solve

$$U_{xx} + U_{yy} = 0, \quad 0 < x < 1, \quad 0 < y < 1$$

$$U_x(0,y) = 0, \quad U_x(1,y) = 0, \quad U_y(x,0) = 0, \quad U_y(x,1) = x(1-x)$$

7.3. Obtain the solution to the following problem using the Monte Carlo method:

$$\nabla^2 U = 0, \quad 0 < x < \pi, \quad 0 < y < \pi$$

$$U_x(0,y) = 0, \quad U(\pi,y) = 2\cos y, \quad U_y(x,0) = 0, \quad U_y(x,\pi) = 0$$

7.4 Apply the Monte Carlo method in solving

$$\nabla^2 U = 0, \quad 0 < x < 1, \quad 0 < y < 1$$

$$U_x(0,y) = 0, \quad U(1,y) = 1, \quad U_y(x,0) = x^2, \quad U_y(x,1) = 0$$

7.5 Solve the following equation using the Monte Carlo method:

$$\nabla^2 U = xy, \quad 0 < x < 1, \quad 0 < y < 1$$

$$U_x(0,y) = 0, \quad U_x(1,y) = 1, \quad U(x,0) = 0, \quad U(x,1) = 10$$

8

Whole Field Computation

*Nothing can stop the man with the right mental attitude from achieving his goal.
Nothing on earth can help the man with the wrong attitude.*

—Thomas Jefferson

8.1 Introduction

There are different kinds of Monte Carlo methods (MCMs). The popular ones are the fixed random walk, the floating random walk, and the Exodus method. These techniques are known for their efficiency in calculating the potential at a single point. However, this major strength is useless when the potential at all points in the solution grid is desired. To overcome this limitation, several techniques have been proposed. These include the shrinking boundary method [1], inscribed figure method [2], and the surface density method [3].

The shrinking boundary method is similar to the classical fixed random walk except that, once the potential at an interior point is calculated, that point is treated as a boundary or absorbing point. This way, the random-walking particles will have more points to terminate their walks, and the walking time is reduced.

The inscribed figure method is based on the concept of subregion calculation. It involves dividing the solution region into standard shapes or inscribing standard shapes into the region. (By "standard shapes" is meant circles, rectangles, squares, triangles, etc., for which Green's function can be obtained analytically or numerically.) Then, an MCM is used for computing potential along the dividing lines between the shapes and the regions that have nonstandard shapes. Standard analytic methods are used to compute the potentials in the subregions.

Both the shrinking boundary method and the inscribed figure method do not make MCMs efficient for whole field calculation. They still require point-by-point calculations and a large number of tests, as standard Monte Carlo techniques do. Also, these techniques are complicated and hard to program. Therefore, they offer no significant advantage over the standard MCMs.

This chapter proposes a simple technique for whole field calculations. The technique basically calculates the transition probabilities using absorbing

Markov chains and gives an exact solution unlike other numerical techniques such as finite difference and finite element methods.

We begin with a very brief discussion of the regular MCM. This background information is needed for the subsequent sections. We then present the absorbing Markov chains and show how they can be used in calculating the potential for the entire region.

8.2 Regular Monte Carlo Method

This section provides the necessary background for the next section, where Markov chain is discussed.

The most popular MCM is the fixed random walk. Suppose that this method is to be applied in solving Laplace's equation

$$\nabla^2 V = 0 \quad \text{in region R} \tag{8.1}$$

subject to the Dirichlet boundary condition

$$V = V_p \text{ on boundary B} \tag{8.2}$$

The region R is divided into a mesh (as in finite difference), as typically shown in Figure 8.1. Equation 8.1 is replaced by its finite difference equivalent as follows:

$$V(x,y) = p_{x+}V(x+\Delta,y) + p_{x-}V(x-\Delta,y) + p_{y+}V(x,y+\Delta) + p_{y-}V(x,y-\Delta) \tag{8.3}$$

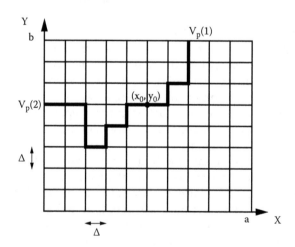

FIGURE 8.1
Fixed random walk from point (x_0, y_0) where the potential is to be determined.

where

$$p_{x+} = p_{x-} = p_{y+} = p_{y-} = \frac{1}{4} \qquad (8.4)$$

A probabilistic interpretation of Equation 8.3 is that if a walking particle is momentarily at point (x,y), the probability of moving to $(x + \Delta, y)$, $(x - \Delta, y)$, $(x, y + \Delta)$, or $(x, y - \Delta)$, is p_{x+}, p_{x-}, p_{y+}, or p_{y-}, respectively.

To find the potential at a free point x_o, y_o (where the potential is unknown), a random-walking particle is asked to begin its walk at that point. It wanders through the mesh according to the probabilities in Equation 8.4 until it reaches the boundary, where it is absorbed and the prescribed potential $V_p(1)$ is recorded. By sending out N particles from x_o, y_o and recording the potential at the end of each walk, we obtain

$$V(x_o, y_o) = \frac{1}{N} \sum_{i=1}^{N} V_p(i) \qquad (8.5)$$

If there are n_p fixed nodes on B with prescribed potential $V_1, V_2, ..., V_{np}$, Equation 8.5 becomes

$$V(x_o, y_o) = \frac{N_1}{N} V_1 + \frac{N_2}{N} V_2 + \cdots + \frac{N_{n_p}}{N} V_{n_p} \qquad (8.6)$$

where N_k is the number of random walks ending at fixed node k and $N_1 + N_2 + \cdots + N_{np} = N$. Thus,

$$V(x_o, y_o) = \sum_{k=1}^{n_p} p_k V_k \qquad (8.7)$$

where $p_k = N_k/N$ is the probability that a random walk starting at free node (x_o, y_o) ends at fixed node k. Calculating p_k is the major goal of this chapter.

In the past, p_k was calculated directly or indirectly just one point (x_o, y_o) at a time using fixed random walk, floating random walk, or the Exodus method. In the next section, we want to employ Markov chains to calculate p_k for all the free nodes in the entire solution region at once.

8.3 Absorbing Markov Chains

A Markov chain is a sequence of random variables $X^{(0)}, X^{(1)}, ..., X^{(n)}$, where the probability distribution of $X^{(n)}$ is determined by the probability distribution $X^{(n-1)}$. A Markov process is a type of random process that is characterized by the memoryless property [4–9]. It is a process evolving in time that remembers only the most recent past and whose conditional probability distributions are time invariant. Markov chains are mathematical models of this kind of process. The Markov chains of interest to us are *discrete-state*, *discrete-time* Markov chains. In our case, the Markov chain is the random walk, and the states are the grid nodes. The transition probability P_{ij} is the probability that a random-walking particle at node i moves to node j. It is expressed by the Markov property

$$P_{ij} = \mathbf{P}(x_{n+1} = j | x_o, x_1, \dots, x_n) = \mathbf{P}(x_{n+1} = j | x_n), \quad j \in X, n = 0, 1, 2, \dots \quad (8.8)$$

The Markov chain is characterized by its transition probability matrix $\mathbf{P}$, defined by

$$\mathbf{P} = \begin{bmatrix} P_{00} & P_{01} & P_{02} & \cdots \\ P_{10} & P_{11} & P_{12} & \cdots \\ P_{20} & P_{21} & P_{22} & \cdots \\ \cdots & \cdots & \cdots & \cdots \end{bmatrix} \quad (8.9)$$

$\mathbf{P}$ is a stochastic matrix, meaning that the sum of the elements in each row is unity, that is,

$$\sum_{j \in X} P_{ij} = 1, \quad i \in X \quad (8.10)$$

We may also use the state transition diagram as a way of representing the evolution of a Markov chain. An example is shown in Figure 8.2 for a three-state Markov chain.

In our case, the Markov chain is the random walk, and the states are the grid nodes. If we assume that there are n_f free (or nonabsorbing) nodes and n_p fixed (prescribed or absorbing) nodes, the size of the transition matrix $\mathbf{P}$ is n, where

$$n = n_f + n_p \quad (8.11)$$

If the absorbing nodes are numbered first and the nonabsorbing states are numbered last, the $n \times n$ transition matrix becomes

$$\mathbf{P} = \begin{bmatrix} 1 & 0 \\ \mathbf{R} & \mathbf{Q} \end{bmatrix} \quad (8.12)$$

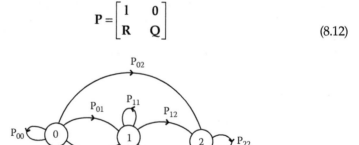

FIGURE 8.2
State transition diagram for a three-state Markov chain.

where the $n_f \times n_p$ matrix $\mathbf{R}$ represents the probabilities of moving from non-absorbing nodes to absorbing ones; the $n_f \times n_f$ matrix $\mathbf{Q}$ represents the probabilities of moving from one nonabsorbing node to another; $\mathbf{I}$ is the identity matrix representing transitions between the absorbing nodes ($P_{ii} = 1$ and $P_{ij} = 0$); and $\mathbf{0}$ is the null matrix showing that there are no transitions from absorbing to nonabsorbing nodes. For the solution of Laplace's equation in Equation 8.1, we obtain the elements of $\mathbf{Q}$ from Equation 8.4 as

$$Q_{ij} = \begin{cases} \dfrac{1}{4}, & \text{if } i \text{ is directly connected to } j \\ 0, & \text{if } i = j \text{ or } i \text{ is not directly connected to } j \end{cases} \tag{8.13}$$

The same applies to R_{ij} except that j is an absorbing node.

For any absorbing Markov chain, $\mathbf{I} - \mathbf{Q}$ has an inverse. This is usually referred to as the *fundamental matrix*

$$\mathbf{N} = (\mathbf{I} - \mathbf{Q})^{-1} \tag{8.14}$$

where N_{ij} is the average number of times the random-walking particle starting from node i passes through node j before being absorbed. The absorption probability matrix $\mathbf{B}$ is

$$\mathbf{B} = \mathbf{N}\,\mathbf{R} \tag{8.15}$$

where R_{ij} is the probability that a random-walking particle originating from a nonabsorbing node i will end up at the absorbing node j. $\mathbf{B}$ is an $n_f \times n_p$ matrix and is stochastic, similar to the transition probability matrix, that is,

$$\sum_{j=1}^{n_p} B_{ij} = 1, \quad i = 1, 2, \ldots, n_f \tag{8.16}$$

If $\mathbf{V}_f$ and $\mathbf{V}_p$ contain potentials at the free and fixed nodes, respectively, then

$$\mathbf{V}_f = \mathbf{B}\mathbf{V}_p \tag{8.17}$$

In terms of the prescribed potentials $V_1, V_2, \ldots, V_{np}$, used in Equations 8.6 and 8.7, Equation 8.17 becomes

$$V_i = \sum_{j=1}^{n_p} B_{ij} V_j, \quad i = 1, 2, \ldots, n_f \tag{8.18}$$

where V_i is the potential at any free node i. Unlike Equation 8.7, Equations 8.17 or 8.18 provide the solution at all the free nodes at once.

An alternative way to obtain the solution in Equation 8.17 is to exploit a property of the transition probability matrix $\mathbf{P}$. When $\mathbf{P}$ is multiplied by itself repeatedly a large number of times, we obtain

$$\lim_{n \to \infty} \mathbf{P}^n = \begin{bmatrix} \mathbf{I} & \mathbf{0} \\ \mathbf{B} & \mathbf{0} \end{bmatrix} \tag{8.19}$$

Thus,

$$
\begin{bmatrix} \mathbf{V}_p \\ \mathbf{V}_f \end{bmatrix} = \mathbf{P}^n \begin{bmatrix} \mathbf{V}_p \\ \mathbf{V}_f \end{bmatrix} = \begin{bmatrix} 1 & 0 \\ \mathbf{B} & 0 \end{bmatrix} \begin{bmatrix} \mathbf{V}_p \\ \mathbf{V}_f \end{bmatrix}
\tag{8.20}
$$

Either Equation 8.17 or Equation 8.20 can be used to find V_f, but it is evident that Equation 8.17 will be more efficient and accurate. From Equation 8.17 or Equation 8.18, it should be noted that if $\mathbf{N}$ is calculated accurately, the solution is "exact."

Three simple examples will corroborate the claims in Section 8.2. The first two are taken from [10] and do not require any computer programming, whereas the last one is taken from [11].

Example 8.1

Consider an infinitely long conducting trough with a square cross section. A conducting lid is maintained at 100 V while the sides and bottom are grounded as shown in Figure 8.3. We wish to determine the potential at the center.

Solution

Mathematically, the problem is posed as follows:

$$
\nabla^2 V = 0
\tag{8.21}
$$

subject to

$$
V(0,y) = V(a,y) = V(x,0), \quad V(x,a) = 100
\tag{8.22}
$$

and we are to determine $V(a/2, a/2)$. Although one may assume that $a = 1$, that is not necessary.

The exact solution obtained by the separation of variables is [12]

$$
V(x,y) = \frac{400}{\pi} \sum_{k=1}^{\infty} \frac{\sin \frac{n\pi x}{a} \sinh \frac{n\pi y}{a}}{n \sinh n\pi}, \quad n = 2k - 1
\tag{8.23}
$$

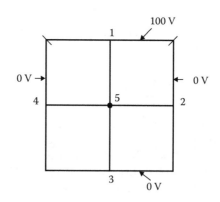

FIGURE 8.3
Figure for Example 8.1.

To apply Markov chain technique, we number the nodes as in Figure 8.3. Node 5 is the only free node $n_f = 1$, $n_p = 4$. The transition probability matrix is given by

$$\mathbf{P} = \begin{array}{c} \\ 1 \\ 2 \\ 3 \\ 4 \\ 5 \end{array} \begin{array}{ccccc} 1 & 2 & 3 & 4 & 5 \\ \begin{bmatrix} 1 & 0 & 0 & 0 & 0 \\ 0 & 1 & 0 & 0 & 0 \\ 0 & 0 & 1 & 0 & 0 \\ 0 & 0 & 0 & 1 & 0 \\ \dfrac{1}{4} & \dfrac{1}{4} & \dfrac{1}{4} & \dfrac{1}{4} & \dfrac{1}{4} \end{bmatrix} \end{array}$$

It is evident that

$$\mathbf{Q} = 0, \quad \mathbf{N} = (\mathbf{1} - \mathbf{Q})^{-1} = \mathbf{I}$$

and

$$\mathbf{R} = \begin{bmatrix} \dfrac{1}{4} & \dfrac{1}{4} & \dfrac{1}{4} & \dfrac{1}{4} \end{bmatrix}$$

Thus,

$$\mathbf{B} = \mathbf{NR} = \begin{bmatrix} \dfrac{1}{4} & \dfrac{1}{4} & \dfrac{1}{4} & \dfrac{1}{4} \end{bmatrix}$$

and

$$\mathbf{V}_f = \mathbf{BV}_p = \begin{bmatrix} \dfrac{1}{4} & \dfrac{1}{4} & \dfrac{1}{4} & \dfrac{1}{4} \end{bmatrix} \begin{bmatrix} V_1 \\ V_2 \\ V_3 \\ V_4 \end{bmatrix}$$

or

$$V_5 = \frac{1}{4}(100 + 0 + 0 + 0) = 25.0$$

which agrees with the exact solution in Equation 8.23. Although the method gives the exact solution in this case, one cannot generalize from this simple example. If the regular fixed random walk method is used and we dispatch 1000 particles from point 5, we may get the numbers of particles absorbed by boundary points as $N_1 = 252$, $N_2 = 251$, $N_3 = 250$, $N_4 = 247$ so that

$$V_5 = \frac{252}{1000}100 + 0 = 25.2$$

which is less accurate than the solution from Markov chain and is subject to randomness.

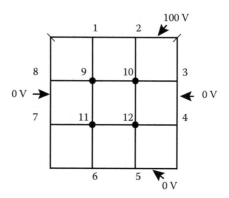

FIGURE 8.4
Figure for Example 8.2.

Example 8.2

This is the same problem as in Example 8.1, except that we are now to calculate the potential at points $(a/3, a/3)$, $(a/3, 2a/3)$, $(2a/3, a/3)$, and $(2a/3, 2a/3)$.

Solution

In this case, there are four free nodes $(n_f = 4)$ and eight fixed nodes $(n_p = 8)$, as shown in Figure 8.4. The transition probability matrix is obtained by inspection as follows:

$$
\mathbf{P} =
\begin{array}{c|cccccccccccc}
 & 1 & 2 & 3 & 4 & 5 & 6 & 7 & 8 & 9 & 10 & 11 & 12 \\
\hline
1 & 1 & & & & & & & & & & & \\
2 & & 1 & & & & & & & & & & \\
3 & & & 1 & & & & & & & & & \\
4 & & & & 1 & & & & & & & & \\
5 & & & & & 1 & & & & & & & \\
6 & & & & & & 1 & & & & & & \\
7 & & & & & & & 1 & & & & & \\
8 & & & & & & & & 1 & & & & \\
9 & \frac{1}{4} & & & & & & & & \frac{1}{4} & 0 & \frac{1}{4} & \frac{1}{4} & 0 \\
10 & 0 & \frac{1}{4} & \frac{1}{4} & & & & & & \frac{1}{4} & 0 & 0 & \frac{1}{4} \\
11 & 0 & 0 & 0 & & & & \frac{1}{4} & \frac{1}{4} & 0 & \frac{1}{4} & & \frac{1}{4} \\
12 & 0 & & & \frac{1}{4} & \frac{1}{4} & & 0 & 0 & 0 & \frac{1}{4} & \frac{1}{4} & 0 \\
\end{array}
$$

Other entries in **P** shown vacant are zeros.

From **P**, we obtain

$$
R = \begin{array}{c} \\ 9 \\ 10 \\ 11 \\ 12 \end{array}
\begin{array}{cccccccc}
1 & 2 & 3 & 4 & 5 & 6 & 7 & 8 \\
\end{array}
\left[
\begin{array}{cccccccc}
\frac{1}{4} & 0 & 0 & 0 & 0 & 0 & 0 & \frac{1}{4} \\
0 & \frac{1}{4} & \frac{1}{4} & 0 & 0 & 0 & 0 & 0 \\
0 & 0 & 0 & 0 & 0 & \frac{1}{4} & \frac{1}{4} & 0 \\
0 & 0 & 0 & \frac{1}{4} & \frac{1}{4} & 0 & 0 & 0 \\
\end{array}
\right]
$$

$$
Q = \begin{array}{c} \\ 9 \\ 10 \\ 11 \\ 12 \end{array}
\begin{array}{cccc}
9 & 10 & 11 & 12 \\
\end{array}
\left[
\begin{array}{cccc}
0 & \frac{1}{4} & \frac{1}{4} & 0 \\
\frac{1}{4} & 0 & 0 & \frac{1}{4} \\
\frac{1}{4} & 0 & 0 & \frac{1}{4} \\
0 & \frac{1}{4} & \frac{1}{4} & 0 \\
\end{array}
\right]
$$

The fundamental matrix N is obtained as follows:

$$
N = (I - Q)^{-1} =
\left[
\begin{array}{cccc}
1 & -\frac{1}{4} & -\frac{1}{4} & 0 \\
-\frac{1}{4} & 1 & 0 & -\frac{1}{4} \\
-\frac{1}{4} & 0 & 1 & -\frac{1}{4} \\
0 & -\frac{1}{4} & -\frac{1}{4} & 1 \\
\end{array}
\right]^{-1}
$$

or

$$
N = \frac{1}{6}
\left[
\begin{array}{cccc}
7 & 2 & 1 & 1 \\
2 & 7 & 1 & 2 \\
2 & 1 & 7 & 2 \\
1 & 2 & 2 & 7 \\
\end{array}
\right]
$$

The absorption probability matrix B is obtained as follows:

$$B = NR = \begin{array}{c} \\ 9 \\ 12 \\ 10 \\ 11 \end{array} \begin{array}{cccccccc} 1 & 2 & 3 & 4 & 5 & 6 & 7 & 8 \\ \left[\begin{array}{cccccccc} \dfrac{7}{24} & \dfrac{1}{12} & \dfrac{1}{12} & \dfrac{1}{24} & \dfrac{1}{24} & \dfrac{1}{12} & \dfrac{1}{12} & \dfrac{7}{24} \\ \dfrac{1}{12} & \dfrac{7}{24} & \dfrac{7}{24} & \dfrac{1}{12} & \dfrac{1}{12} & \dfrac{1}{24} & \dfrac{1}{24} & \dfrac{1}{12} \\ \dfrac{1}{12} & \dfrac{1}{24} & \dfrac{1}{24} & \dfrac{1}{12} & \dfrac{1}{12} & \dfrac{7}{24} & \dfrac{7}{24} & \dfrac{1}{12} \\ \dfrac{1}{24} & \dfrac{1}{12} & \dfrac{1}{12} & \dfrac{7}{24} & \dfrac{7}{24} & \dfrac{1}{12} & \dfrac{1}{12} & \dfrac{1}{24} \end{array} \right] \end{array}$$

Note that Equation 8.16 is satisfied. We now use Equation 8.18 to obtain the potentials at the free nodes. For example,

$$V_9 = \frac{7}{24}V_1 + \frac{1}{12}V_2 + \frac{1}{12}V_3 + \frac{1}{24}V_4 + \frac{1}{24}V_5 + \frac{1}{12}V_6 + \frac{1}{12}V_7 + \frac{1}{24}V_8$$

Because $V_1 = V_2 = 100$ and $V_3 = V_4 = \cdots = V_8 = 0$,

$$V_9 = \left(\frac{7}{24} + \frac{1}{12} \right) 100 = 37.5$$

By symmetry, $V_{10} = V_9 = 37.5$. Similarly,

$$V_{11} = V_{12} = \left(\frac{1}{24} + \frac{1}{12} \right) 100 = 12.5$$

Table 8.1 compares these results with the finite difference solution (with 10 iterations) and the exact solution using Equation 8.23. It is evident that the Markov chain solution compares well.

TABLE 8.1

Results of Example 8.2

Node	Finite Difference Solution	Markov Chain Solution	Exact Solution
9	38.499	38.5	38.074
10	38.499	38.5	38.074
11	12.499	12.5	11.926
12	12.499	12.5	11.926

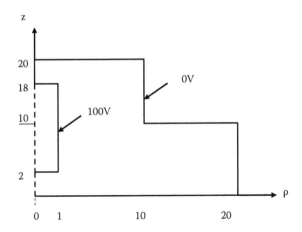

FIGURE 8.5
Figure for Example 8.3.

TABLE 8.2

Results of Example 8.3

| Node | | Markov | Exodus | Finite |
ρ	z	Chain	Method	Difference
5	18	11.3931	11.438	11.474
5	10	28.4727	28.816	28.869
5	2	12.3099	12.179	12.128
10	2	2.4481	2.3523	2.3421
15	2	0.4684	0.38423	0.3965

Example 8.3

The last example is an axisymmetric problem shown in Figure 8.5. The problem is a prototype of an electrostatic particle-focusing system used in a recoil-mass time-of-flight spectrometer. Using a step size of $\Delta = 0.01$, we obtain the results in Table 8.2. The problem has no exact solution, and we compare the results from Markov chain with those from Exodus and finite difference methods at five typical points in Table 8.2. It is clear from the table that the results from the three methods are close.

8.4 Summary

This chapter has presented a means of using MCM to solve Laplace's equation for the entire solution region at once as opposed to a single-point calculation. The approach is using Markov chains to calculate the transition probabilities. This approach is not subject to randomness because a random number generator is not required. The approach also provides an exact solution. The ideas presented in this chapter can be extended to solution regions that are inhomogeneous or nonrectangular or both [11]. All it takes is calculating the transition probability **P**. The idea of Markov chain may also be used to solve Poisson's equation and the wave equation.

References

1. G. E. Zinsmeiter and S. S. Pan, A method for improving the efficiency of Monte Carlo calculation of heat conduction problems, *J. Heat Transfer, Trans. ASME, Series C*, vol. 96, 246–248, 1974.
2. G. E. Zinsmeiter and S. S. Pan, A modification of the Monte Carlo method, *Inter. J. Num. Meth. Engr.*, vol. 10, 1057–1064, 1976.
3. T. J. Hoffman and N. E. Banks, Monte Carlo solution to the Dirichlet problem with the double-layer potential density, *Trans. Amer. Nucl. Sci.*, vol. 18, 1974, pp. 136,138. See also vol. 19, 1974, p. 164; vol. 24. 1976, p. 181.
4. Fusco, V. F. and Linden, P. A., A Markov chain approach for static field analysis, *Microwave Optical Technol. Lett.*, vol. 1, no. 6, August 1988, pp. 216–220.
5. M. E. Woodward, *Communication and Computer Networks*, Los Alamitos, CA: IEEE Computer Society Press, 1994, pp. 53–58.
6. J. G. Kemeny and J. L. Snell, *Finite Markov Chains*, New York: Springer-Verlag, 1976, pp. 43–68.
7. M. Iosifescu, *Finite Markov Processes and Their Applications*, New York: John Wiley & Sons, 1980, pp. 45, 99–106.
8. G. J. Anders, *Probability Concepts in Electric Power Systems*, New York: John Wiley & Sons, 1990, pp. 160–170.
9. L. Bonet-Cunha et al., A hybrid Markov chain Monte Carlo method for generating permeability fields conditioned to multiwell pressure data and prior information, *SPE J.*, vol. 3, no. 3, 261–271, September 1998.
10. M. N. O. Sadiku, K. Gu, and C. N. Obiozor, Regional Monte Carlo potential calculation using Markov chains, to appear in *Int. J. Engineering Education*.
11. M. N. O. Sadiku and R. C. Garcia, Whole field computation using Monte Carlo method, *Int. J. Numer. Model.*, vol. 10, 303–312, 1997.
12. M. N. O. Sadiku, *Elements of Electromagnetics*, 4th ed., New York: Oxford University Press, 2007, pp. 221–229.

Problems

8.1. Consider the potential problem shown in Figure 8.6. Typically, use the values

$$V_o = 100, \quad \varepsilon_1 = \varepsilon_o, \quad \varepsilon_2 = 2.25\varepsilon_o, \quad a = 3.0, \quad b = 2.0, \quad c = 1.0$$

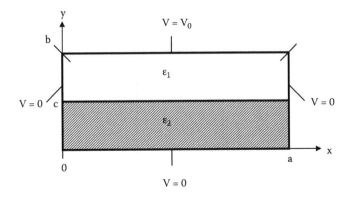

FIGURE 8.6
Figure for Problem 8.1.

Calculate the potential at points $(x,y) = (0.5, 1.0)$, $(1.0, 1.0)$, $(1.5, 1.0)$, $(1.5, 0.5)$, $(1.5, 1.5)$ using Markov chain.

8.2. For the potential system shown in Figure 8.7, take

$$V_o = 100, \quad \varepsilon_1 = \varepsilon_o, \quad \varepsilon_2 = 3\varepsilon_o, \quad a = b = 0.5, \quad h = w = 1.0$$

At the corner point $(x,y) = (a,b)$, the random walk probabilities are

$$p_{x+} = p_{x-} = \frac{\varepsilon_1}{3\varepsilon_1 + \varepsilon_2}, \quad p_{y+} = p_{y-} = \frac{\varepsilon_1 + \varepsilon_2}{2(3\varepsilon_1 + \varepsilon_2)}$$

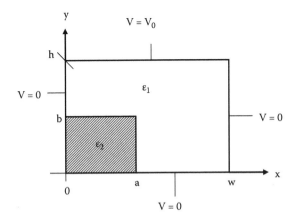

FIGURE 8.7
Figure for Problem 8.2.

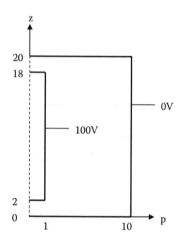

FIGURE 8.8
Figure for Problem 8.3.

Using Markov chains, determine the potential at points $(x, y) =$ (0.25, 0.5), (0.5, 0.5), (0.75, 0.5), (0.5, 0.75), and (0.5, 0.25).

8.3. Consider the finite cylindrical conductor held at $V = 100$ enclosed in a larger grounded cylinder, as portrayed in Figure 8.8. Using Markov chain, write a program to determine the potential at points $(\rho, z) = (2, 10), (5, 10), (8, 10), (5, 2),$ and $(5, 18)$.

9

Time-Varying Problems

From birth to eighteen, a girl needs good parents. From eighteen to thirty-five, she needs good looks. From thirty-five to fifty-five, good personality. From fifty-five on, she needs good cash.

—Sophie Tucker

9.1 Introduction

Monte Carlo Methods (MCMs) have been applied successfully for solving differential and integral equations, finding eigenvalues, inverting matrices, and evaluating multiple integrals. MCMs are well known for solving static problems such as Laplace's or Poisson's equations. They are hardly applied in solving parabolic and hyperbolic partial differential equations. In this chapter, we extend the applicability of the conventional MCM to the solution of time-dependent (heat) problems. We deal with cases of rectangular solution regions as well as axisymmetric solution regions. We present results in one dimension (1-D) and two dimensions (2-D) that agree with the finite difference (FD) and exact solutions. Most of the information presented in this chapter is taken from [1–4].

9.2 Diffusion Equation

Consider the skin effect on a solid cylindrical conductor. The current density distribution within a good conducting wire ($\sigma/\omega \in \gg 1$) obeys the diffusion equation

$$\nabla^2 \mathbf{J} = \mu\sigma \frac{\partial \mathbf{J}}{\partial t} \tag{9.1}$$

We may derive the diffusion equation directly from Maxwell's equation. We recall that

$$\nabla \times \mathbf{H} = \mathbf{J} + \mathbf{J}_d \tag{9.2}$$

where $J = \sigma E$ is the conduction current density and $J_d = \frac{\partial D}{\partial t}$ is the displacement current density. For $\sigma/\omega\, \epsilon \gg 1$, J_d is negligibly small compared to J. Hence,

$$\nabla \times H \approx J \tag{9.3}$$

Also,

$$\nabla \times E = -\mu\frac{\partial H}{\partial t}$$

$$\nabla \times \nabla \times E = \nabla\nabla \cdot E - \nabla^2 E = -\mu\frac{\partial}{\partial t}\nabla \times H \tag{9.4}$$

Because $\nabla \cdot E = 0$, introducing Equation 9.3, we obtain

$$\nabla^2 E = \mu\frac{\partial J}{\partial t} \tag{9.5}$$

Replacing E with J/σ, Equation 9.5 becomes

$$\nabla^2 J = \mu\sigma\frac{\partial J}{\partial t}$$

which is the diffusion (or heat) equation.

9.3 Rectangular Solution Region

We now consider the Monte Carlo solution of diffusion in 1-D and 2-D forms in the rectangular coordinate system.

9.3.1 One-Dimensional Case

Consider the 1-D heat equation:

$$U_{xx} = U_t, \quad 0 < x < 1, \quad t > 0 \tag{9.6}$$

Boundary conditions:

$$U(0,t) = 0 = U(1,t), \quad t > 0 \tag{9.7a}$$

Initial condition:

$$U(x,0) = 100, \quad 0 < x < 1 \tag{9.7b}$$

In Equation 9.6, U_{xx} indicates the second partial derivative with respect to x, whereas U_t indicates the partial derivative with respect to t. The problem models temperature distribution in a rod or eddy current in a conducting medium [1]. To solve this problem using MCM, we first need to obtain the FD equivalent of the partial differential equation (Equation 9.6). Using the central-space and backward-time scheme, we obtain

$$\frac{U(i+1,n)-2U(i,n)+U(i-1,n)}{(\Delta x)^2} = \frac{U(i,n)-U(i,n-1)}{\Delta t} \qquad (9.8)$$

where $x = i\Delta x$ and $t = n\Delta t$. If we let

$$\alpha = \frac{(\Delta x)^2}{\Delta t} \qquad (9.9)$$

Equation 9.8 becomes

$$U(i,n) = p_{x+}U(i+1,n) + p_{x-}U(i-1,n) + p_{t-}U(i,n-1) \qquad (9.10)$$

where

$$p_{x+} = p_{x-} = \frac{1}{2+\alpha}, \quad p_{t-} = \frac{\alpha}{2+\alpha} \qquad (9.11)$$

Note that $p_{x+} + p_{x-} + p_{t-} = 1$. Equation 9.10 can be given a probabilistic interpretation. If a random-walking particle is instantaneously at the point (x,y), it has probabilities p_+, p_-, and p_{t-} of moving from (x, t) to $(x + \Delta x, t)$, $(x - \Delta x, t)$, and $(x, t - \Delta t)$, respectively. The particle can only move toward the past, never toward the future. A means of determining which way the particle should move is to generate a random number r, $0 < r < 1$, and instruct the particle to walk as follows:

$$(x,t) \longrightarrow (x+\Delta x,t) \quad \text{if } (0 < r < 0.25)$$

$$(x,t) \longrightarrow (x-\Delta x,t) \quad \text{if } (0.25 < r < 0.5)$$

$$(x,t) \longrightarrow (x,t-\Delta t) \quad \text{if } (0.5 < r < 1) \qquad (9.12)$$

where it is assumed that $\alpha = 2$. Most modern software packages such as MATLAB have a random number generator to obtain r.

To calculate U at point (x_o, t_o), we follow the following random walk algorithm:

1. Begin a random walk at $(x, t) = (x_o, t_o)$.
2. Generate a random number $0 < r < 1$, and move to the next point using Equation 9.12.
3a. If the next point is not on the boundary, repeat step 2.

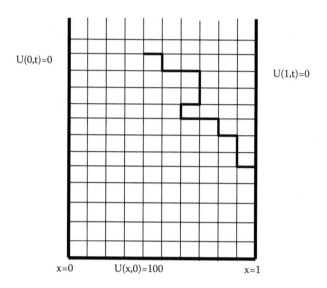

FIGURE 9.1
A typical random walk in the rectangular domain.

3b. If the random walk hits the boundary, terminate the random walk. Record U_b at the boundary and go to step 1 and begin another random walk.

4. After N random walks, determine

$$U(x_o, t_o) = \frac{1}{N} \sum_{k=1}^{N} U_b(k)$$

(9.13)

where N, the number of random walks, is assumed to be large.

A typical random walk is illustrated in Figure 9.1.

Example 9.1

As a numerical example, consider the solution of the problem in Equations 9.6 and 9.7. We select $\alpha = 2$, $\Delta x = 0.1$, so that $\Delta t = 0.005$ and $p_{x+} = p_{x-} = \frac{1}{4}$, $p_{t-} = \frac{1}{2}$.

Solution

We calculate U at $x_o = 0.4$, $t = 0.01, 0.02, 0.03, \ldots$. The MATLAB code for the problem is shown in Figure 9.2. As shown in Table 9.1, we compare the results with the FD and exact solutions [5]:

$$U(x, t) = \frac{400}{\pi} \sum_{k=0}^{\infty} \frac{1}{n} \sin(n\pi x) \exp(-n^2 \pi^2 t), \quad n = 2k + 1$$

(9.14)

```
% This program solves one-dimensional diffusion (or heat) equation
% using Monte Carlo method

nrun = 1000;
delta = 0.1;
% deltat=2*delta^2;
deltat = 0.005;
A=1.0;
xo=0.4;
to=0.1;
io=xo/delta;
jo=to/deltat;
no=to/deltat;
imax=A/delta;
sum=0;
for k=1:nrun
    i=io;
    n=no;
    while i<=imax & n<=no
    r=rand; %random number between 0 and 1
    if (r >= 0.0 & r <= 0.25)
        i=i+1;
    end
    if (r >= 0.25 & r <= 0.5)
        i=i-1;
     end
     if (r  >= 0.5 & r <= 1.0)
        n=n-1;
     end
```

(a)

FIGURE 9.2
MATLAB program for Example 9.1.

```
if (n < 0)

    break;

end

% check if (i,n) is on the boundary

    if(i == 0.0)

      sum=sum+ 0.0;

        break;

    end

    if(i == imax)

      sum=sum+ 0.0;

      break;

    end

    if(n == 0.0)

      sum=sum+ 100;

      break;

    end

  end % while


    end

u=sum/nrun
```

(b)

FIGURE 9.2
(Continued).

TABLE 9.1

Comparing Monte Carlo (MCM)
Solution with Finite Difference
(FD) and Exact Solutions ($x_o = 0.4$)

t	Exact	MCM	FD
0.01	99.53	99.44	100
0.02	95.18	93.96	96.87
0.03	89.32	87.62	89.84
0.04	80.88	81.54	82.03
0.10	45.13	46.36	45.18

9.3.2 Two-Dimensional Case

Suppose we are interested in the solution of the 2-D heat equation:

$$U_{xx} + U_{yy} = U_t, \quad 0 < x < 1, \ 0 < y < 1, \ t > 0 \tag{9.15}$$

Boundary conditions:

$$U(0, y, t) = 0 = U(1, y, t), \quad 0 < y < 1, t > 0 \tag{9.16a}$$

$$U(x, 0, t) = 0 = U(x, 1, t), \quad 0 < x < 1, t > 0 \tag{9.16b}$$

Initial condition:

$$U(x, y, 0) = 10xy, \quad 0 < x < 1, \ 0 < y < 1 \tag{9.16c}$$

Using the central-space and backward-time scheme, we obtain the FD equivalent as

$$\frac{U(i+1, j, n) - 2U(i, j, n) + U(i-1, j, n)}{(\Delta x)^2} + \frac{U(i, j+1, n) - 2U(i, j, n) + U(i, j-1, n)}{(\Delta y)^2}$$

$$= \frac{U(i, j, n) - U(i, j, n-1)}{\Delta t} \tag{9.17}$$

Let $\Delta x = \Delta y = \Delta$ and

$$\alpha = \frac{\Delta^2}{\Delta t} \tag{9.18}$$

Equation 9.17 becomes

$$U(i, j, n) = p_{x+} U(i+1, j, n) + p_{x-} U(i-1, j, n)$$

$$+ p_{y+} U(i, j+1, n) + p_{y-} U(i, j-1, n)$$

$$+ p_{t-} U(i, j, n-1) \tag{9.19}$$

where

$$p_{x+} = p_{x-} = p_{y+} = p_{y-} = \frac{1}{4+\alpha} \tag{9.20a}$$

$$p_{t-} = \frac{\alpha}{4+\alpha} \tag{9.20b}$$

Note that $p_{x+} + p_{x-} + p_{y+} + p_{y-} + p_{t-} = 1$, so a probabilistic interpretation can be given to Equation 9.19. A random-walking particle at point (x, y, t) moves to $(x + \Delta, y, t)$, $(x - \Delta, y, t)$, $(x, y + \Delta, t)$, $(x, y - \Delta, t)$, or $(x, y, t - \Delta t)$ with probabilities

$p_{x+}, p_{x-}, p_{y+}, p_{y-},$ or $p_{t-},$ respectively. By generating a random number $0 < r < 1,$ we instruct the particle to move as follows:

$$(x,y,t) \longrightarrow (x+\Delta, y, t) \quad \text{if } (0 < r < 0.2)$$

$$(x,y,t) \longrightarrow (x-\Delta, y, t) \quad \text{if } (0.2 < r < 0.4)$$

$$(x,y,t) \longrightarrow (x, y+\Delta, t) \quad \text{if } (0.4 < r < 0.6)$$

$$(x,y,t) \longrightarrow (x, y-\Delta, t) \quad \text{if } (0.6 < r < 0.8)$$

$$(x,y,t) \longrightarrow (x, y, t-\Delta t) \quad \text{if } (0.8 < r < 1) \tag{9.21}$$

assuming that $\alpha = 1.$ Therefore, we take the following steps to calculate U at point (x_o, y_o, t_o):

1. Begin a random walk at $(x, y, t) = (x_o, y_o, t_o)$.
2. Generate a random number $0 < r < 1,$ and move the next point according to Equation 9.21.
3a. If the next point is not on the boundary, repeat step 2.
3b. If the random walk hits the boundary, terminate the random walk. Record U_b at the boundary and go to step 1 and begin another random walk.
4. After N random walks, determine

$$U(x_o, y_o, t_o) = \frac{1}{N} \sum_{k=1}^{N} U_b(k) \tag{9.22}$$

The only difference between 1-D and 2-D is that there are three kinds of displacements in 1-D whereas there are five displacements (four spatial ones and one temporal one) in 2-D.

Example 9.2

As a numerical example, consider the solution of the problem in Equations 9.15 and 9.16.

Solution

We select $\alpha = 1,$ $\Delta = 0.1,$ so that $\Delta t = 0.01,$ and we calculate U at $x = 0.5,$ $y = 0.5,$ $t = 0.05, 0.1, 0.15, 0.2, 0.25, 0.3.$ As shown in Table 9.2, we compare the results from the MCM with the FD solution and exact solution [5]:

$$U(x,y,t) = \frac{40}{\pi^2} \sum_{m=1}^{\infty} \sum_{n=1}^{\infty} \frac{\cos(m\pi)\cos(n\pi)}{mn} \sin(m\pi x)$$

$$\times \sin(n\pi y) \exp\left(-\lambda_{mn}^2 t\right), \tag{9.23}$$

where $\lambda_{mn}^2 = (m\pi)^2 + (n\pi)^2.$

TABLE 9.2

Comparing Monte Carlo Solution
with Finite Difference and Exact
Solutions ($x_o = 0.5$, $y_o = 0.5$)

t	Exact	MCM	FD
0.05	1.491	1.534	1.518
0.10	0.563	0.6627	0.5627
0.15	0.216	0.267	0.2063
0.20	0.078	0.106	0.0756
0.25	0.029	0.0419	0.0277
0.30	0.015	0.019	0.0102

Because of the randomness of the Monte Carlo solution, each MCM result in Tables 9.1 and 9.2 was obtained by running the simulation five times and taking the average.

9.4 Cylindrical Solution Region

We now consider applying the Monte Carlo method to 1-D and 2-D cases of cylindrical solution regions.

9.4.1 One-Dimensional Case

Consider the 1-D heat equation in cylindrical coordinates:

$$\nabla^2 U = \frac{\partial U}{\partial t} \tag{9.24a}$$

Or

$$U_{\rho\rho} + \frac{1}{\rho} U_\rho = U_t, \quad 0 < \rho < 1, \quad t > 0 \tag{9.24b}$$

Boundary conditions:

$$U(1,t) = 0, \quad t > 0 \tag{9.25a}$$

Initial condition:

$$U(\rho,0) = T_o \tag{9.25b}$$

The problem models temperature distribution in a thin circular plate. To solve this problem using MCM, we first need to obtain the FD equivalent of the partial differential equation in Equation 9.24. Using the central-space and backward-time scheme, we obtain

$$\frac{U(i+1,n)-2U(i,n)+U(i-1,n)}{(\Delta\rho)^2} + \frac{U(i+1,n)-U(i-1,n)}{\rho 2\Delta\rho}$$

$$= \frac{U(i,n)-U(i,n-1)}{\Delta t} \tag{9.26}$$

where $\rho = i\Delta\rho = ih$ and $t = n\Delta t$. If we let

$$\alpha = \frac{h^2}{\Delta t} \tag{9.27}$$

Equation 9.26 becomes

$$U(i,n) = p_{\rho+}U(i+1,n) + p_{\rho-}U(i-1,n) + p_{t-}U(i,n-1) \tag{9.28}$$

where

$$p_{\rho+} = \frac{1+1/2i}{2+\alpha}, \quad p_{\rho-} = \frac{1-1/2i}{2+\alpha}, \quad p_{t-} = \frac{\alpha}{2+\alpha} \tag{9.29}$$

Note that $p_{\rho+} + p_{\rho-} + p_{t-} = 1$. Equation 9.28 can be given a probabilistic interpretation. If a random-walking particle is instantaneously at the point (ρ,t), it has probabilities $p_{\rho+}, p_{\rho-}$, and p_{t-} of moving from (ρ,t) to $(\rho+h,t)$, $(\rho-h,t)$, and $(\rho, t-\Delta t)$, respectively. The particle can only move toward the past, but never toward the future. A means of determining which way the particle should move is to generate a random number r, $0 < r < 1$, and instruct the particle to walk as follows:

$$(\rho,t) \longrightarrow (\rho+h,t) \quad \text{if } (0 < r < prho)$$

$$(\rho,t) \longrightarrow (\rho-h,t) \quad \text{if } (prho < r < 0.5)$$

$$(\rho,t) \longrightarrow (\rho,t-\Delta t) \quad \text{if } (0.5 < r < 1) \tag{9.30}$$

where $prho = (1+1/(2i))/(2+\alpha)$ and it is assumed that $\alpha = 2$. Most modern software packages such as MATLAB® have a random number generator to obtain r.

Equations 9.26 to 9.30 are applicable for the case $\rho \neq 0$. For $\rho = 0$, we need to apply L'Hopital's rule to Equation 9.24.

$$\lim_{\rho \to 0} \frac{1}{\rho}\frac{\partial U}{\partial \rho} = \frac{\partial^2 U}{\partial \rho^2} \tag{9.31}$$

Thus,

$$\nabla^2 U = \frac{\partial U}{\partial t} \rightarrow 2U_{\rho\rho} = U_t \qquad (9.32)$$

Applying the central-space and backward-time scheme to Equation 9.32 and noting that $U(h, t) = U(-h, t)$, we obtain

$$U(0,t) = p_{\rho+} U(h,t) + p_{t-} U(0,t - \Delta t) \qquad (9.33)$$

where

$$p_{\rho+} = \frac{4}{4+\alpha}, \quad p_{t-} = \frac{\alpha}{4+\alpha} \qquad (9.34)$$

When a random-walking particle is at $\rho = 0$, a means of determining which way the particle should move is to generate a random number r, $0 < r < 1$, and instruct the particle to walk as follows:

$$(0,t) \longrightarrow (h,t) \quad \text{if } (0 < r < prhop)$$

$$(0,t) \longrightarrow (0,t - \Delta t) \quad \text{if } (prhop < r < 1) \qquad (9.35)$$

where $prhop = 4/(4+\alpha)$ and it is assumed that $\alpha = 2$.

To calculate U at point (ρ_o, t_o), we follow the following random walk algorithm:

1. Begin a random walk at $(\rho, t) = (\rho_o, t_o)$.
2. Generate a random number $0 < r < 1$, and move to the next point using Equation 9.35 if $\rho = 0$ or Equation 9.30 if $\rho \neq 0$.
3a. If the next point is not on the boundary, repeat step 2.
3b. If the random walk hits the boundary, terminate the random walk. Record U_b at the boundary, go to step 1, and begin another random walk.
4. After N random walks, determine

$$U(\rho_o, t_o) = \frac{1}{N} \sum_{k=1}^{N} U_b(k) \qquad (9.36)$$

where N, the number of random walks, is assumed to be large.

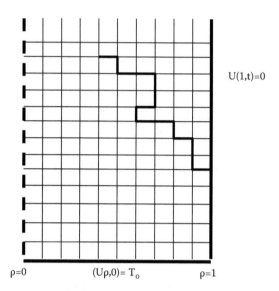

FIGURE 9.3
A typical random walk in cylindrical domain.

A typical random walk for the problem is illustrated in Figure 9.3. Note that $\rho = 0$ is not a boundary but a line of symmetry.

Example 9.3
As a numerical example, consider the solution of the problem in Equations 9.24 and 9.25.

Solution

We select $T_o = 10, \alpha = 2, h = 0.1$, so that $\Delta t = 0.005$. We calculate U at $\rho = 0.5$ and $t = 0.1, 0.2, 0.3, ...$ As shown in Table 9.3, we compare the results with the

TABLE 9.3

Comparing Monte Carlo
(MCM) Solution with Finite
Difference (FD) and Exact
Solutions ($\rho_o = 0.5$)

t	Exact	MCM	FD
0.1	6.0191	6.0640	6.0573
0.2	3.3752	3.3920	3.3532
0.3	1.8933	1.9340	1.8770
0.4	1.0619	1.0880	1.0516
0.5	0.5955	0.5880	0.5893
1.0	0.0330	0.0320	0.0325

finite difference solution and exact solution [6]:

$$U(\rho, t) = 2T_0 \sum_{n=1}^{\infty} \frac{J_0(\lambda_n \rho)}{\lambda_n J_1(\lambda_n)} \exp\left(-\lambda_n^2 t\right),$$ (9.37)

where J_0 and J_1 are Bessel functions of order 0 and 1, respectively, and the values of λ_n are the positive roots of J_0.

9.4.2 Two-Dimensional Case

Suppose we are interested in the solution of the 2-D heat equation:

$$U_{\rho\rho} + \frac{U_\rho}{\rho} + U_{zz} = U_t, \quad 0 < \rho < 1, \, 0 < z < 1, \, t > 0$$ (9.38)

Boundary conditions:

$$U(\rho, 0, t) = 0 = U(\rho, 1, t), \quad 0 < \rho < 1, \, t > 0$$ (9.39a)

$$U(1, z, t) = 0, \quad 0 < z < 1, \, t > 0$$ (9.39b)

Initial condition:

$$U(\rho, z, 0) = T_0, \, 0 < \rho < 1, \, 0 < z < 1$$ (9.39c)

This models the temperature distribution in a solid cylinder of unit height and unit radius.

Using the central-space and backward-time scheme, we obtain the FD equivalent as

$$\frac{U(i+1, j, n) - 2U(i, j, n) + U(i-1, j, n)}{(\Delta\rho)^2} + \frac{U(i+1, j, n) - U(i-1, j, n)}{\rho 2\Delta\rho}$$

$$+ \frac{U(i, j+1, n) - 2U(i, j, n) + U(i, j-1, n)}{(\Delta z)^2} = \frac{U(i, j, n) - U(i, j, n-1)}{\Delta t}$$ (9.40)

Let $\Delta x = \Delta z = h$ and $\rho = ih, z = jh, t = n\Delta t$:

$$\alpha = \frac{h^2}{\Delta t}$$ (9.41)

Equation 9.40 becomes

$$U(i,j,n) = p_{\rho+}U(i+1,j,n) + p_{\rho-}U(i-1,j,n)$$

$$+ p_{z+}U(i,j+1,n) + p_{z-}U(i,j-1,n)$$

$$+ p_{t-}U(i,j,n-1) \tag{9.42}$$

where

$$p_{\rho+} = \frac{1+1/2i}{4+\alpha}, \quad p_{\rho-} = \frac{1-1/2i}{4+\alpha} \tag{9.43a}$$

$$p_{z+} = \frac{1}{4+\alpha}, \quad p_{z-} = \frac{1}{4+\alpha}, \quad p_{t-} = \frac{\alpha}{4+\alpha} \tag{9.43b}$$

Note that $p_{\rho+} + p_{\rho-} + p_{z+} + p_{z-} + p_{t-} = 1$, so a probabilistic interpretation can be given to Equation 9.42. A random-walking particle at point (ρ, z, t) moves to $(\rho+h, z, t)$, $(\rho-h, z, t)$, $(\rho, z+h, t)$, $(\rho, z-h, t)$, or $(\rho, z, t-\Delta t)$ with probabilities $p_{\rho+}, p_{\rho-}, p_{z+}, p_{z-}$, or p_{t-}, respectively. By generating a random number $0 < r < 1$, we instruct the particle to move as follows:

$$(\rho, z, t) \longrightarrow (\rho+h, z, t) \quad \text{if } (0 < r < prho)$$

$$(\rho, z, t) \longrightarrow (\rho-h, z, t) \quad \text{if } (prho < r < 0.4)$$

$$(\rho, z, t) \longrightarrow (\rho, z+h, t) \quad \text{if } (0.4 < r < 0.6)$$

$$(\rho, z, t) \longrightarrow (\rho, z-h, t) \quad \text{if } (0.6 < r < 0.8)$$

$$(\rho, z, t) \longrightarrow (\rho, z, t-\Delta t) \quad \text{if } (0.8 < r < 1) \tag{9.44}$$

assuming that $\alpha = 1$ and $prho = 0.2 * (1 + 1/(2 * i))$.

Equations 9.40 to 9.44 only apply for $\rho \neq 0$. For $\rho = 0$, we apply L'Hopital's rule in Equation 9.38 and obtain

$$2U_{\rho\rho} + U_{zz} = U_t \tag{9.45}$$

We now apply central-space and backward-time scheme to Equation 9.45 and note that $U(h, z, t) = U(-h, z, t)$. We obtain

$$U(0,j,n) = p_{\rho+}U(1,j,n) + p_{z+}U(0,j+1,n) + p_{z-}U(0,j-1,n)$$

$$+ p_{t-}U(0,j,n-1) \tag{9.46}$$

where

$$p_{\rho+} = \frac{4}{6+\alpha}, \quad p_{z+} = p_{z-} = \frac{1}{6+\alpha}, \quad p_{t-} = \frac{\alpha}{6+\alpha} \tag{9.47}$$

A random-walking particle that finds itself at $\rho = 0$ determines the next location by generating a random number r, $0 < r < 1$, and walking as follows:

$$(0,z,t) \longrightarrow (h,z,t) \quad \text{if } (0 < r < 4*pp)$$

$$(0,z,t) \longrightarrow (0,z+h,t) \quad \text{if } (4*pp < r < 5*pp)$$

$$(0,z,t) \longrightarrow (0,z-h,t) \quad \text{if } (5*pp < r < 6*pp)$$

$$(0,z,t) \longrightarrow (0,z,t-\Delta t) \quad \text{if } (6*pp < r < 1) \qquad (9.48)$$

where $pp = 1/(6+\alpha)$ and it is assumed that $\alpha = 1$.

Therefore, we take the following steps to calculate U at point (ρ_o, z_o, t_o):

1. Begin a random walk at $(\rho, z, t) = (\rho_o, z_o, t_o)$.
2. Generate a random number $0 < r < 1$, and move the next point according to Equation 9.44 if $\rho \neq 0$ or Equation 9.48 if $\rho = 0$.
3a. If the next point is not on the boundary, repeat step 2.
3b. If the random walk hits the boundary, terminate the random walk. Record U_b at the boundary, go to step 1, and begin another random walk.
4. After N random walks, determine

$$U(\rho_o, z_o, t_o) = \frac{1}{N} \sum_{k=1}^{N} U_b(k) \qquad (9.49)$$

The only difference between 1-D and 2-D is that there are three kinds of displacement in 1-D whereas there are five displacements (four spatial ones and one temporal one) in 2-D.

Example 9.4

As a numerical example, consider the solution of the problem in Equations 9.38 and 9.39.

Solution

We select $\alpha = 1$, $T_o = 10$, and $h = 0.1$, so that $\Delta t = 0.01$ and we calculate U at $\rho = 0.5$, $z = 0.5$, $t = 0.05$, 0.1, 0.15, 0.2, 0.25, 0.3. Figure 9.4 has the MATLAB program for the problem. As shown in Table 9.4, we compare the results from the MCM with the FD solution and exact solution [7,8]:

$$U(\rho,z,t) = \frac{8T_o}{\pi} \sum_{m=1}^{\infty} \sum_{n=1,3,5}^{\infty} \frac{J_o(k_m \rho)}{nk_m J_1(k_m)} \sin(n\pi z) \exp\left(-\lambda_{mn}^2 t\right) \qquad (9.50)$$

where $\lambda_{mn}^2 = k_m^2 + (n\pi)^2$ and k_m is the m-th root of the Bessel function $J_o(k_m)$.

```
% This program solves two-dimensional diffusion (or heat) equation

% in cylindrical, i.e. U_rhorho + U_rho/rho + U_zz = U_t

% using Monte Carlo method


nrun = 1000;

h = 0.1;

% deltat=delta^2;

deltat = 0.01;

alpha = 1.0;

To=10;

A=1.0;

B=1.0;

rho=0.5;

zo=0.5;

to=0.1;

io=rho/h;

jo=zo/h;

no=to/deltat;

imax=A/h;

jmax=B/h;

sum2=0.0;

for m=1:5

    sum=0;

for k=1:nrun

   i=io;

   j=jo;

   n=no;

   while i<=imax & j<=jmax & n<=no
```

FIGURE 9.4
MATLAB code for Example 9.4.

```
       if(i == 0.0)  % this is along the line of symmetry

           pp = 1/(6 + alpha);

        r=rand; %random number between 0 and 1

        if (r >= 0.0 & r <= 4*pp)

           i=i+1;

        end

         if (r >= 4*pp & r <= 5*pp)

          j=j+1;

        end

       if (r >= 5*pp & r <= 6*pp)

          j=j-1;

         end

        if (r  >= 6*pp & r <= 1.0)

            n=n-1;

         end

    else

            prhoplus = 0.2*(1 + 1/(2*i));

        r=rand; %random number between 0 and 1

        if (r >= 0.0 & r <= prhoplus)

           i=i+1;

        end

       if (r >= prhoplus & r <= 0.4)

          i=i-1;

         end

        if (r >= 0.4 & r <= 0.6)

          j=j+1;

        end

       if (r >= 0.6 & r <= 0.8)

          j=j-1;
```

FIGURE 9.4
(Continued).

```
        end
        if (r  >= 0.8 & r <= 1.0)

            n=n-1;

        end

    end

    if (n < 0)

        break;

    end

    % check if (i,j,n) is on the boundary

    if(i == imax)

        sum=sum+ 0.0;

        break;

    end

     if(j == 0.0)

        sum=sum+ 0.0;

            break;

    end

    if(j == jmax)

        sum=sum+ 0.0;

        break;

    end

     if(n == 0.0)

        sum=sum + To;

        break;

    end

  end % while

 end

uu=sum/nrun

sum2 = sum2 + uu;

end

u=sum2/5
```

FIGURE 9.4
(Continued).

TABLE 9.4

Comparing Monte Carlo Solution
with Finite Difference and Exact
Solutions ($\rho_0 = 0.5$, $z_0 = 0.5$)

t	Exact	MCM	FD
0.05	6.2475	6.614	6.3848
0.10	2.8564	3.182	2.9123
0.15	1.3059	1.582	1.2975
0.20	0.5971	0.7760	0.5913
0.25	0.2730	0.4140	0.270
0.30	0.1248	0.156	0.1233

Because of the randomness of the Monte Carlo solution, each MCM result in Tables 9.1 to 9.4 was obtained by running the simulation five times and taking the average.

9.5 Summary

In this chapter, we have demonstrated how the conventional MCM can be applied to time-dependent problems such as the heat equation in both rectangular and cylindrical coordinates. For 1-D and 2-D cases, we note that the Monte Carlo solutions agree well with the FD and the exact analytic solutions, and it is easier to understand and program than the FD method. The method does not require solving large matrices and is trivially easy to program. The idea can be extended to other time-dependent problems such as the wave equation.

References

1. D. Netter, J. Levenque, P. Masson, and A. Rezzoug, Monte Carlo method for transient eddy-current calculations, *IEEE Trans. Magnetics*, vol. 40, no. 5, pp. 3450–3456, September 2004.

2. M. N. O. Sadiku, C. M. Akujuobi, and S. M. Musa, Monte Carlo analysis of time-dependent problems, *Proc. of IEEE Southeastcon*, 2006, pp. 7–10.

3. M. N. O. Sadiku, C. M. Akujuobi, S. M. Musa, and S. Nelatury, Monte Carlo analysis of time-dependent cylindrical problems, *Proc. of IEEE Southeastcon*, 2007, pp. 778–782.

4. M. N. O. Sadiku, C. M. Akujuobi, S. M. Musa, and S. R. Nelatury, Analysis of time-dependent cylindrical problems using Monte Carlo, *Microwave Optical Technol. Lett.*, vol. 49, no. 10, pp. 2571–2573, October 2007.

5 M. N. O. Sadiku, *Numerical Techniques in Electromagnetics*, 2nd ed., Boca Raton, FL: CRC Press, 2001, pp. 125–130.

6. D. L. Powers, *Boundary Value Problems*, New York: Academic Press, 1972, pp. 135–139.

7. L. C. Andrews, *Elementary Partial Differential Equations with Boundary Value Problems*, Orlando, FL: Academic Press, 1986, pp. 459, 466, 467.

8. L. C. Andrews, *Answer Booklet to accompany Elementary Partial Differential Equations with Boundary Value Problems*, Orlando, FL: Academic Press, 1986, p. 33.

Problems

9.1. Use Monte Carlo method to solve the heat equation described by

$$kU_{xx} = U_t, \quad 0 \leq x \leq 1, t > 0$$

$$U(0,t) = 0, \quad t > 0,$$

$$U(1,t) = 1, \quad t > 0$$

$$U(x,0) = 0, \quad 0 \leq x \leq 1$$

9.2. Using Monte Carlo method, find $U(x,t)$ satisfying

$$U_t = U_{xx}, \quad 0 < x < 2, \quad t > 0$$

$$U(0,t) = 0, \quad U(2,t) = 0, \quad t > 0$$

$$U(x,0) = \begin{cases} 2x, & 0 < x < 0.5 \\ 1, & 0.5 < x < 1.5 \\ 4 - 2x, & 1.5 < x < 2 \end{cases}$$

9.3. Use Monte Carlo method to solve

$$\frac{\partial^2 U}{\partial x^2} = \frac{\partial U}{\partial t}, \quad 0 < x < 1, \quad t > 0$$

$$U(0,t) = 10, \quad t > 0$$

$$U(1,t) = 20, \quad t > 0$$

$$U(x,0) = 0, \quad 0 < x < 1$$

9.4. Find $U(x,t)$ that satisfies the following four conditions using Monte Carlo method:

$$U_t = U_{xx}, \quad 0 < x < 1, \quad 0 < t < \infty$$

$$U(0,t) = 0, \quad 0 < t < \infty$$

$$U(1,t) = 0, \quad 0 < t < \infty$$

$$U(x,0) = 10 \sin \pi x, \quad 0 < x < 1$$

9.5. Solve the following heat equation using Monte Carlo techniques:

$$U_{xx} + U_{yy} = U_t, \quad 0 < x < \pi, 0 < y < \pi$$

$$U(0,y,t) = 0, \quad U(\pi,y,t) = 0$$

$$U(x,0,t) = 0, \quad U(x,\pi,t) = 0$$

$$U(x,y,0) = 10$$

10

Scattering from Random Rough Surfaces

The man without a purpose is like a ship without a rudder—a waif, a nothing, a no man.

—Thomas Carlyle

10.1 Introduction

Several natural surfaces are rough in varying degrees, and this roughness affects the propagation and scattering of electromagnetic waves [1]. Scattering of EM waves from random rough surfaces has been a subject of interest for a long time. Monte Carlo simulations of wave scattering by random rough surfaces have been performed since the 1970s. The problem has been studied extensively owing to its many practical applications such as microwave scattering by ocean or terrain, ultrasound scattering by tissues, and underwater acoustics. Rough surfaces are also boundaries for active remote sensing.

The classical analytic approaches of Kirchhoff approximation and Rayleigh–Rice small perturbation method [2–5] have been used to solve the problem of random rough surface scattering. However, both are restricted in domain validity. With the advent of modern computers, the simulations of random rough surfaces have been performed. The general procedure involves forming an integral equation in the space domain. The integral equation is converted to a matrix equation by the method of moments, and the resulting equation is then solved with a full matrix inversion. Recently, finite difference and finite element methods have been used for the Monte Carlo simulation of random rough surfaces.

Because the treatment of random rough surface in this chapter is meant to be introductory, we begin with a one-dimensional (1-D) rough surface (two-dimensional [2-D] scattering problem). We later discuss EM wave scattering by a 2-D random rough surface.

10.2   Scattering by 1-D Random Rough Surfaces

The first step in constructing a 1-D random rough surface is to generate an uncorrelated Gaussian distribution of random numbers $f(x)$, where x is the coordinate along the surface and f is the corresponding random height [6]. This is achieved by considering

$$f(x) = \sum_{i=1}^{M} U_j(x) - M/2 \tag{10.1}$$

where $U_j(x)$ is an equally distributed random number in the interval $(0,1)$ with expectation value and correlation

$$<U_j(x)> = 0.5 \tag{10.2a}$$

$$<[U_i(x)-0.5][U_j(y)-0.5]> = \frac{1}{12}\delta_{ij}\delta_{xy} \tag{10.2b}$$

so that

$$< f(x) > = 0 \tag{10.3a}$$

$$< f(x)f(y) > = \frac{1}{12} M\delta_{xy} \tag{10.4}$$

According to the central limit theorem, $f(x)$ is Gaussian distributed for large M. In practice, it is common to select $M = 12$. This leads to very small deviations in the tails of the distribution.

We now consider a tapered time-harmonic plane wave $\Psi^{inc}(x,z)$ with time dependence $\exp(-i\omega t)$ impinging on a 1-D rough surface with a random height profile $z = f(x)$; see Figure 10.1. The incident wave is tapered so that the illuminated surface can be confined to length L.

$$\Psi^{inc}(r) = e^{ik\bullet r(1+W(r))}e^{-(x+z\tan\theta_{inc})^2/g^2} \tag{10.5}$$

where

$$k = k(\sin\theta_{inc}a_x - \cos\theta_{inc}a_z) \tag{10.6a}$$

$$W(r) = \frac{[2(x+z\tan\theta_{inc})^2/g^2 - 1]}{(kg\cos\theta_{inc})^2} \tag{10.6b}$$

Parameter g controls the tapering of the wave, and its value satisfied $1/(kg \cos\theta_{inc}) \le 0.037$ for numerical simulation [7–14]. The boundary condition requires that the wave function $\Psi = 0$ on the rough surface. This constitutes

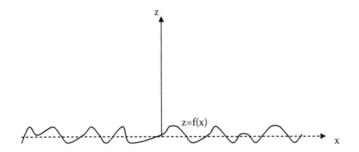

FIGURE 10.1
One-dimensional random rough surface.

the Dirichlet problem implying a TE electromagnetic wave impinging on a perfectly conducting rough surface and is also known as the *s-wave case*. We can form the Fredholm integral equation of the first kind. For a point $xa_x + f(x)a_z$ on the rough surface, the integral equation is

$$0 = \Psi^{inc}(r) - \int dx' \sqrt{1 + \left(\frac{df(x')}{dx'}\right)^2} \, G(r,r') \frac{\partial \Psi(r')}{\partial n'} \qquad (10.7)$$

where

$$G(r,r') = \frac{i}{4} H_0^{(1)}(k|r-r'|) \qquad (10.8)$$

is the free-space Green's function and $H_0^{(1)}$ is the Hankel function of the first kind.

By discretizing the space domain in the 1-D rough surface, Equation 10.7 can be cast into a matrix equation,

$$[Z][X] = [C] \qquad (10.9)$$

where $[X]$ is a column vector of dimension N representing the unknown values of

$$U(x) = \sqrt{1 + (df/dx)^2} \, \partial \Psi / \partial n \qquad (10.10)$$

and $[C]$ is a column vector representing the first term on the right-hand side of Equation 10.7. The matrix element Z_{mn} of matrix $[Z]$ denotes the interaction between two points on the rough surface. When Equation 10.9 is solved directly, it is called the *exact matrix inversion* (EMI). Other approaches to solving the matrix equation include conjugate gradient method (CGM) and banded matrix iterative approach (BMIA). In the application of method of

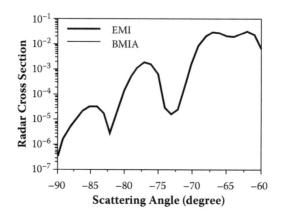

FIGURE 10.2
Comparison of the BMIA and EMI for one realization. (From L. Tsang et al., *J. Electromagnetic Waves and Applications*, vol. 7, no. 9, 1185–1200, 1993. With permission.)

moments to rough surface scattering problem, a common approach is to use a grid of 10 points per wavelength to discretize the surface.

Once the solution of Equation 10.9 is obtained, we can get the radar cross section as follows:

$$\sigma(\theta_s) = \frac{\frac{1}{8\pi} \left| \int_{-\infty}^{\infty} dx e^{-ik(\sin\theta_s x + \cos\theta_s f(x))} U(x) \right|^2}{gk\sqrt{\frac{\pi}{2}} \cos\theta_{inc} \left(1 - \frac{1 + 2\tan^2\theta_{inc}}{2k^2 g^2 \cos^2\theta_{inc}} \right)} \tag{10.11}$$

where θ_s is the scattering angle and $U(x)$ is defined in Equation 10.10. Figure 10.2 shows the radar cross section when the surface length $L = 100\lambda$, rms height $h = 0.575\lambda$, at an angle of incidence $\theta_{inc} = 80°$, and with 800 unknowns for one realization [14]. The tapering parameter g is set at $L/4$.

10.3 Scattering by 2-D Random Rough Surfaces

We now consider the case of three-dimensional (3-D) problems with 2-D random rough surfaces [15–22]. Using a Monte Carlo simulation, we first generate many sample surfaces of a stochastic random process representing the desired random rough surface. A typical rough surface profile is generated and shown in Figure 10.3. The random surface height is defined by function $z = f(x, y)$ with rms surface height h and correlation length l in both x and y directions.

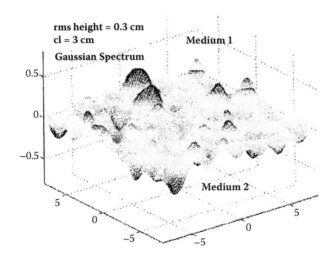

FIGURE 10.3
Generated 2-D surface profile with rms height of 0.3 cm and correlation length of 3 cm. (From K. W. Lam et al., *IEEE Trans. on Antennas and Propagation*, vol. 32, no. 12, 3200–3206, December 2004. With permission.)

Consider a scalar plane wave $\Psi_{inc}(x,y,z)$ with time dependence $\exp(-i\omega t)$ impinging on a 2-D perfectly conducting rough surface with a random height profile $z = f(x, y)$. The incident direction is $k_i = \sin\theta_i \cos\phi_i\, a_x + \sin\theta_i \sin\phi_i a_y - \cos\theta_i a_z$. The incident wave is tapered so that the illuminated rough surface can be confined to the surface area $L \times L$. It is given by

$$\Psi_{inc}(x,y,z) = \exp[-ik(\cos\theta_i z - x\sin\theta_i \cos\phi_i - y\sin\theta_i \sin\phi_i)(1+\omega_o)]\exp(-d_x - d_y)$$

$$(10.12)$$

where

$$\omega_o = \frac{1}{k^2}\left[\frac{(2d_x - 1)}{g^2 \cos^2\theta_i} + \frac{(2d_y - 1)}{g^2}\right] \qquad (10.13)$$

$$d_x = \frac{(\cos\theta_i \cos\phi_i x + \cos\theta_i \sin\phi_i y + \sin\theta_i z)^2}{g^2 \cos^2\theta_i} \qquad (10.14)$$

$$d_y = \frac{(-\sin\phi_i x + \cos\phi_i y)^2}{g^2} \qquad (10.15)$$

$$k = \omega\sqrt{\mu_o\varepsilon_o} = 2\pi/\lambda \qquad (10.16)$$

The parameter g controls the tapering of the incident wave.

The Dirichlet boundary condition is $\Psi(x, y, f(x, y)) = 0$ on the rough surface $z = f(x, y)$. We can develop a Fredholm integral of the first kind. Let $r = xa_x + ya_y + za_z$ represent the source point on the rough surface and $r' = x'a_x + y'a_y + z'a_z$ represent the field point. The integral equation is

$$0 = \Psi_{inc}(r) - \iint dxdy G_o[x, y, f(x, y); x', y', f(x', y')] U(x, y) \qquad (10.17)$$

where

$$G_o(r, r') = \frac{\exp(ik|r - r'|)}{4\pi|r - r'|} \qquad (10.18)$$

is the free-space Green's function. The unknown surface variable $U(x, y)$ is given as follows:

$$U(x) = \frac{\partial \Psi(r)}{\partial n} \left[1 + \left(\frac{\partial f}{\partial x} \right)^2 + \left(\frac{\partial f}{\partial y} \right)^2 \right]^{1/2} \qquad (10.19)$$

The integration equation (Equation 10.17) can be solved by the moment method and other approaches. The Monte Carlo simulations require that the computation be repeated for many rough surface realizations until results converge. Solving a 3-D scattering problem is expensive even for one surface realization, so conducting Monte Carlo simulations with hundreds of realizations without using fast algorithms is prohibitive even with supercomputers.

After the unknown $U(x, y)$ is determined numerically either by using the method of moments or the conjugate gradient method, the normalized bistatic scattering coefficient $\sigma(k_s)$ in the direction of k_s is calculated as

$$\sigma(k_s) = \frac{|F(k_s)|^2}{8\pi^3 g^2 \cos\theta_i \left[1 - \frac{(1 + \cos^2\theta_i + 2\tan^2\theta_i)}{2k^2 g^2 \cos^2\theta_i} \right]} \qquad (10.20)$$

where

$$F(k_s) = -\iint dxdy U(x, y) \exp[-ikx\sin\theta_s \cos\theta_s - iky\sin\theta_s \sin\phi_s - ikf(x, y)\cos\theta_s] $$

$$(10.21)$$

The normalized bistatic scattering coefficient $\sigma(k_s)$ is defined such that its integration over the 2π solid angle of the upper-half space equals unity.

Figure 10.4 depicts the normalized bistatic scattering coefficient for the angle of incidence of $\theta_i = 20°$ and $\phi_i = 0°$ and tapering parameter $g = L/3$. Four cases are shown, for 155, 225, 275, and 310 realizations. The cases of 275 and 310 realizations overlap each other, indicating that the coefficient converges in approximately 310 realizations [15].

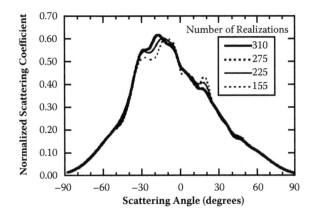

FIGURE 10.4
Normalized bistatic scattering coefficient versus the scattering angle. (From L. Tsang, C. H. Chan, and K. Pak, *J. Optical Society of America*, vol. 11, no. 2, 711–715, February 1994. With permission.)

10.4 Summary

The study of wave scattering by random rough surfaces is a topic of continued interest because of its broad applications. In this chapter, we have considered 1-D and 2-D perfectly conducting random rough surfaces. The idea can be extended to dielectric rough surfaces [23–25]. Although the method used in this chapter is the tapered wave integral equation formulation and its solution by the method of moments, the problem has also been solved using a finite-difference time-domain (FDTD) approach [26] and finite element method [8, 27–30]. The benefits of using FDTD include the following: (1) matrix inversion is eliminated, (2) time-domain response is readily available, and (3) scattering intensities can be obtained simultaneously for a range of frequencies. The attraction of the finite element method is the banded nature of the resulting matrix equation. All the nonzero elements of the matrix are confined to a band or profile. For a symmetric matrix, only the storage of the elements in the upper half of the band is necessary. The method yields a system of linear algebraic equations that can be solved using a direct sparse symmetric matrix inversion.

References

1. A. Ishimaru, *Wave Propagation and Scattering in Random Media*, vol. 2, New York: Academic Press, 1978, p. 463.
2. P. Beckmann and A. Spizzichino, *The Scattering of Electromagnetic Waves from Rough Surfaces*, New York: Pergamon, 1963.

3. A. Ishimaru, *Wave Propagation and Scattering in Random Media*, San Diego, CA: Academic, 1978.

4. L. Tsang, J. A. Jong, and R. T. Shin, *Theory of Microwave Remote Sensing*, New York: Wiley & Sons, 1985.

5. E. I. Thorsos, The validity of the Kirchhoff approximation for rough surface scattering using a Gaussian roughness spectrum, *J. Acoustic Society of America*, vol. 83, no. 1, 78–92, January 1988.

6. N. Garcia and E. Stoll, Monte Carlo calculations for electromagnetic-wave scattering from random rough surfaces, *Physical Review Lett.*, vol. 52, no. 20, 1798–1801, May 1984.

7. L. Tsang et al., Monte Carlo simulations of large-scale composite random rough-surface scattering based on the banded-matrix iterative approach, *J. Optical Society of America, A: Optics, Image Science, and Vision*, vol. 11, no. 2, 691–696, February 1994.

8. K. C. Krause et al., Application of the finite element method to Monte Carlo simulations of random rough surface scattering with Neumann boundary conditions, *Microwave Optical Technol. Lett.*, vol. 4, no. 7, 255–258, June 1991.

9. L. Tsang et al., Monte-Carlo simulations of large-scale problems of random rough surface scattering and applications to gracing incidence with BMIA/canonical grid method, *IEEE Trans. Antennas and Propagation*, vol. 43, no. 8, 851–859, August 1995.

10. Q. Li, C. H. Chan, and L. Tsang, Monte Carlo simulations of wave scattering from lossy dielectric random rough surfaces using the physics-based two-grid method and the canonical-grid method, *IEEE Trans. Antennas and Propagation*, vol. 47, no. 4, 752–763, April 1999.

11. L. Tsang, C. H. Chan, and H. Sangani, Application of a banded matrix iterative approach to Monte Carlo simulations of scattering of waves by a random rough surface: TM case, *Microwave Optical Technol. Lett.*, vol. 6, no. 2, 148–151, 1993.

12. L. Tsang, C. H. Chan, and H. Sangani, Banded matrix iterative approach to Monte-Carlo simulations of scattering of waves by large scale random rough surface problems: TM case, *Electronic Lett.*, vol. 29, No, 2, 166–167, January 1993.

13. M. Nieto-Vesperinas and J. M. Soto-Crespo, Monte Carlo simulations for scattering of electromagnetic waves from perfectly conductive random rough surfaces, *Optics Lett.*, vol. 12, no. 12, 979–981, December 1987.

14. L. Tsang et al., A banded matrix iterative approach to Monte Carlo simulations of large-scale random rough surface scattering: TE case, *J. Electromagnetic Waves and Applications*, vol. 7, no. 9, 1185–1200, 1993.

15. K. W. Lam et al., On the analysis of statistical distributions of UWB signal scattering by random rough surfaces based on Monte Carlo simulations of Maxwell's equations, *IEEE Trans. on Antennas and Propagation*, vol. 32, no. 12, 3200–3206, December 2004.

16. K. Pak, L. Tsang, and C. H. Chan, Backscattering enhancement of electromagnetic waves from two-dimensional perfectly conducting random rough surfaces on Monte Carlo simulations, *J. Optical Society of America*, vol. 12, no. 11, 2491–2499, November 1995.

17. L. Tsang, C. H. Chan, and K. Pak, Backscattering enhancement of a two-dimensional random rough surface (three-dimensional scattering) based on Monte Carlo simulations, *J. Optical Society of America*, vol. 11, no. 2, 711–715, February 1994.

18. J. T. Johnson et al., Backscattering enhancement of electromagnetic waves from two-dimensional perfectly conducting random rough surfaces: a comparison of Monte Carlo simulations with experimental data, *IEEE Trans. on Antennas and Propagation,* vol. 44, no. 5, 748–756, May 1996.

19. L. Tsang, C. H. Chan, and K. Pak, Monte Carlo simulation of a two-dimensional random rough surface using the sparse-matrix flat-surface iterative approach, *Electronics Lett.,* vol. 29, no. 13, 1153–1154, June 1993.

20. R. L. Wagner, J. Song, and W. C. Chew, Monte Carlo simulation of electromagnetic scattering from two-dimensional random rough surfaces, *IEEE Trans. on Antennas and Propagation,* vol. 45, no. 2, 235–245, February 1997.

21. L. Tsang et al. (Eds.), *Scattering of Electromagnetic Waves: Numerical Simulations,* New York: John Wiley & Sons, 2001, pp. 267–370.

22. L. Tsang, S. H. Lou, and C. H. Chan, Application of the extended boundary condition method to Monte Carlo simulations of scattering of waves by two-dimensional random rough surfaces, *Microwave Optical Technol. Lett.,* vol. 4, no. 12, 527–531, November 1991.

23. M. Sallard and D. Maystra, Scattering from metallic and dielectric rough surfaces, *J. Optical Society of America, A: Optics, Image Science, and Vision,* vol. 2, no. 6, 982–990, June 1990.

24. C. H. Chan, L. Tsang, and Q. Li, Monte Carlo simulations of large-scale one-dimensional random rough-surface scattering at near-grazing incidence: penetrable case, *IEEE Trans. on Antennas and Propagation,* vol. 46, no.1, 142–1410, January 1998.

25. M. El-Shenawee, C. Rappaport, and M. Silevitch, Monte Carlo simulations of electromagnetic wave scattering from a random rough surface with three-dimensional penetrable buried objects: mine detection application using the steepest-descent fast multipole method, *J. Optical Society of America, A: Optics, Image Science, and Vision,* vol. 18, no. 12, 3077–3084, December 2001.

26. C. H. Chan et al., Electromagnetic scattering of waves by random rough surface: a finite-difference time-domain approach, *Microwave and Optical Technol. Lett.,* vol. 4, no. 9, 355–359, August 1991.

27. S. H. Lou et al., Application of the finite element method to Monte Carlo simulations of scattering of waves by random rough surfaces with the periodic boundary condition, *J. Electromagnetic Waves and Applications,* vol. 5, no. 8, 835–855, 1991.

28. S. H. Lou, L. Tsang, and C. H. Chan, Application of the finite element method to Monte Carlo simulations of scattering of waves by random rough surfaces: penetrable case, *Waves in Random Media,* vol. 1, 287–307, 1991.

29. L. Li et al., Monte Carlo simulations and backscattering enhancement of random metallic rough surfaces at optical frequencies, *J. Electromagnetic Waves and Applications,* vol. 8, no. 3, 277–293, 1994.

30. K. Pak, Combined random rough surfaces and volume scattering based on Monte Carlo simulations of solutions of Maxwell's equations, *Radio Science,* vol. 28, no. 3, May–June 1990, pp. 331–338.

11

Multidimensional Integration

Experience is not what happens to you. It is what you do with what happens to you.

—Aldous Huxley

11.1 Introduction

One can use integral calculus to find the area under a curve, provided the curve has a closed form. When the curve has no closed form, the area cannot be derived analytically and one can employ Monte Carlo method (MCM) to calculate the area.

Also, for one-dimensional integration, several quadrature formulas (such as trapezoidal rule, Simpson's rule, etc.) are available. The numbers of such formulas are relatively few for multidimensional integration. It is for such multidimensional integrals that a Monte Carlo technique becomes valuable, for at least two reasons. The quadrature formulas become very complex for multiple integrals, whereas the MCM formula remains almost unchanged. The convergence of Monte Carlo integration is independent of dimensionality, which is not true of quadrature formulas.

The statistical method of integration has been found to be an efficient way to evaluate two- or three-dimensional integrals in antenna problems, particularly those involving very large structures [1]. Two types of Monte Carlo integration procedures, the crude MCM and the MCM with antithetic variates, will be discussed. For other types, such as hit-or-miss and control variates, see [2–4]. Application of MCM to improper integrals will be covered briefly.

11.2 Crude Monte Carlo Integration

Let us consider a circle inscribed in a square on a dartboard, as shown in Figure 11.1. Finding the area of the square is relatively easy, but it is much harder to find the area of the circle. Suppose we randomly throw a large number of darts at the board and count the number (n) of darts that land

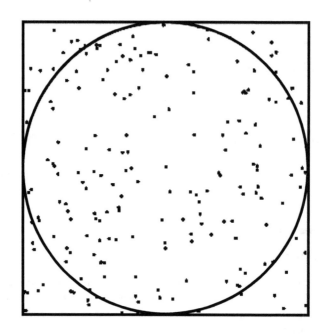

FIGURE 11.1
Finding the area of a circle. (From G. Yang, A Monte Carlo method of integration, http://unicast.org/enclosures/text.pdf. With permission.)

within the square and the number (n^*) that land inside the circle. If A is the area of the square, the area of the circle is [5]

$$I \approx \frac{n^*}{n} A \tag{11.1}$$

Thus, the circle's area is approximately equal to the area of the square multiplied by the fraction of darts that landed inside the circle. We now apply this simple idea.

Suppose we wish to evaluate the integral

$$I = \int_R f \tag{11.2}$$

where R is an n-dimensional space. Let $\mathbf{X} = (X^1, X^2, \ldots, X^n)$ be a random variable that is uniformly distributed in R. Then, $f(\mathbf{X})$ is a random variable whose mean value is given by [6,7]

$$\overline{f(\mathbf{X})} = \frac{1}{|R|} \int_R f = \frac{I}{|R|} \tag{11.3}$$

And the variance by

$$\text{Var}(f(\mathbf{X})) = \frac{1}{|R|}\int_R f^2 - \left(\frac{1}{|R|}\int_R f\right)^2 \tag{11.4}$$

where

$$|R| = \int_R d\mathbf{X} \tag{11.5}$$

If we take N independent samples of $\mathbf{X}$, that is, $(X_1, X_2, \cdots, X_n)$, all having the same distribution as $\mathbf{X}$ and form the average

$$\frac{f(\mathbf{X}_1) + f(\mathbf{X}_2) + \cdots + f(\mathbf{X}_N)}{N} = \frac{1}{N}\sum_{i=1}^{N} f(\mathbf{X}_i) \tag{11.6}$$

We might expect this average to be close to the mean of $f(\mathbf{X})$. Thus, from Equations 11.3 and 11.6,

$$I = \frac{|R|}{N}\sum_{i=1}^{N} f(\mathbf{X}_i) \tag{11.7}$$

This Monte Carlo formula applies to any integration over a finite region R. For the purpose of illustration, we now apply Equation 11.7 to one- and two-dimensional integrals.

For a one-dimensional integral, suppose

$$I = \int_a^b f(x)dx \tag{11.8}$$

Applying Equation 11.7 yields

$$I = \frac{b-a}{N}\sum_{i=1}^{N} f(X_i) \tag{11.9}$$

where X_i is a random number in the interval (a, b), that is,

$$X_i = a + (b-a)U, \quad 0 < U < 1 \tag{11.10}$$

For a two-dimensional integral

$$I = \int_a^b\int_c^d f(X^1, X^2)dX^1 dX^2 \tag{11.11}$$

the corresponding Monte Carlo formula is

$$I = \frac{(b-a)(d-c)}{N} \sum_{i=1}^{N} f\left(X_i^1, X_i^2\right)$$

(11.12)

where

$$X_i^1 = a + (b-a)U^1, \quad 0 < U^1 < 1$$

$$X_i^2 = c + (d-c)U^2, \quad 0 < U^2 < 1$$

(11.13)

The convergence behavior of the unbiased estimator I in Equation 11.7 is slow because the variance of the estimator is of the order $1/N$. Accuracy and convergence are increased by reducing the variance of the estimator using an improved method, the method of antithetic variates. This will be discussed in Section 11.3.

Example 11.1

Evaluate the two-dimensional integral

$$I = \int_0^1 \int_0^1 (x^2 + y^2) dx dy$$

Solution

Although this is a trivial example that can be solved analytically, a numerical integral using Monte Carlo method with 5000 samples gives a value of 0.6664, whereas the exact value is 2/3.

11.3 Monte Carlo Integration with Antithetic Variates

The term *antithetic variates* [8,9] is used to describe any set of estimators that mutually compensate each other's variations. For convenience, we assume that the integral is over the interval (0, 1). Suppose we want an estimator for the single integral

$$I = \int_0^1 g(U) dU$$

(11.14)

We expect the quantity $\frac{1}{2}[g(U) + g(1-U)]$ to have smaller variance than $g(U)$. If $g(U)$ is too small, then $g(1 - U)$ will have a good chance of being too large, and conversely. Therefore, we define the estimator

$$I = \frac{1}{N} \sum_{i=1}^{N} \frac{1}{2} [g(U_i) + g(1 - U_i)] \tag{11.15}$$

where U_i are random numbers between 0 and 1. The variance of the estimators is of the order $\frac{1}{N^4}$, a tremendous improvement over Equation 11.7. For the two-dimensional integral,

$$I = \int_0^1 \int_0^1 g(U^1, U^2) dU^1 dU^2 \tag{11.16}$$

and the corresponding estimator is

$$I = \frac{1}{N} \sum_{i=1}^{N} \frac{1}{4} \Big[g\big(U_i^1, U_i^2\big) + g\big(U_i^1, 1 - U_i^2\big)$$

$$+ g\big(1 - U_i^1, U_i^2\big) + g\big(1 - U_i^1, 1 - U_i^2\big) \Big] \tag{11.17}$$

Following similar lines, the idea can be extended to higher-order integrals. For intervals other than (0, 1), transformations such as in Equations 11.9–11.13 should be applied. For example,

$$\int_a^b f(x) dx = (b - a) \int_0^1 g(U) dU$$

$$\simeq \frac{b - a}{N} \sum_{i=1}^{N} \frac{1}{2} [g(U_i) + g(1 - U_i)] \tag{11.18}$$

where $g(U) = f(X)$ and $X = a + (b - a)U$. It is observed from Equations 11.15 and 11.17 that as the number of dimensions increases, the minimum number of antithetic variates per dimension required to obtain an increase in efficiency over crude Monte Carlo also increases. Thus, the crude Monte Carlo method becomes preferable in many dimensions.

11.4 Improper Integrals

The integral

$$I = \int_0^\infty g(x) dx \tag{11.19}$$

may be evaluated using Monte Carlo simulations [10]. For a random variable X having probability density function $f(x)$, where $f(x)$ integrates to 1 on interval $(0, \infty)$,

$$\int_0^\infty \frac{g(x)}{f(x)} dx = \int_0^\infty g(x)dx \tag{11.20}$$

Hence, to compute I in Equation 11.19, we generate N independent random variables distributed according to a probability density function $f(x)$ integrating to 1 on the interval $(0, \infty)$. The sample mean

$$\overline{g(x)} = \frac{1}{N} \sum_{i=1}^N \frac{g(x_i)}{f(x_i)} \tag{11.21}$$

gives an estimate for I.

Example 11.2

Evaluate the integral

$$I = \int_0^1 \int_0^{2\pi} e^{j\alpha\rho\cos\phi} \rho\, d\rho\, d\phi \tag{11.22}$$

using the Monte Carlo method.

Solution

This integral represents radiation from a circular aperture antenna with a constant amplitude and phase distribution. It is selected because it forms at least part of every radiation integral. The solution is available in the closed form, which can be used to assess the accuracy of the Monte Carlo results. In closed form,

$$I(\alpha) = \frac{2\pi J_1(\alpha)}{\alpha} \tag{11.23}$$

where $J_1(\alpha)$ is a Bessel function of the first order.

A simple program for evaluating the integral employing Equations 11.12 and 11.17, where $a = 0$, $b = 1$, $c = 0$, and $d = 2\pi$, is shown in Figure 11.2. The main program (*integral.m*) calls the random number generator *rand* to generate random numbers U^1 and U^2. It also uses a function (*fun.m*), which is kept in a separate file and shown in Figure 11.2b. For different values of N, both the crude and antithetic variate Monte Carlo methods are used for evaluating the radiation integral, and the results are compared with the exact value in Table 11.1 for $\alpha = 5$. In applying Equation 11.17, the following correspondences are used:

$$U^1 \equiv X^1, \quad U^2 \equiv X^2, \quad 1 - U^1 \equiv b - X^1 = (b-a)(1-U^1),$$

$$1 - U^2 \equiv d - X^2 = (d-c)(1-U^2)$$

```
% Integration using crude Monte Carlo
% and antithetic methods
%
% Only few lines need be changed to use this
% program for any multi-dimensional integration
%
% the function fun.m is to be on a separate file


a =0; b= 1.0; c = 0;  % limits of integration
d=2*pi;
alpha = 5;
nrun = 10000;
sum1 = 0; sum2 = 0;
for i=1:nrun
    u1 = rand;
    u2 = rand;
    x1 = a + (b-a)*u1;
    x2 = c + (d-c)*u2;
    x3 = (b-a)*(1-u1);
    x4 = (d-c)*(1-u2);
    sum1 = sum1 + fun(x1,x2);
    sum2 = sum2 + fun(x1,x2) + fun(x1,x4) + fun(x3,x2) + fun(x3,x4);
end
area1 = (b-a)*(d-c)*sum1/nrun
area2 = (b-a)*(d-c)*sum2/(4*nrun)
                              (a)
```

```
function y=fun(rho,phi)
alpha = 5;
y=rho*exp(j*alpha*rho*cos(phi));
```

(b)

FIGURE 11.2
MATLAB program for Monte Carlo evaluation of a two-dimensional integral: (a) main program, (b) function *fun.m* kept in a separate file.

TABLE 11.1

Results of Example 11.2 on Monte Carlo Integration of Radiation Integral

N	Crude MCM	Antithetic variates MCM
500	$-0.2892 - j0.0742$	$-0.2887 - j0.0585$
1000	$-0.5737 + j0.0808$	$-0.4982 - j0.0080$
2000	$-0.4922 - j0.0040$	$-0.4682 - j0.0082$
4000	$-0.3999 - j0.0345$	$-0.4216 - j0.0323$
6000	$-0.3608 - j0.0270$	$-0.3787 - j0.0440$
8000	$-0.4327 - j0.0378$	$-0.4139 - j0.0241$
10,000	$-0.4229 - j0.0237$	$-0.4121 - j0.0240$

Exact: $-0.4116 + j0$

11.5 Summary

Monte Carlo integration is numerical quadrature using random numbers. In other words, Monte Carlo integration methods are algorithms for the approximate evaluation of definite integrals. The regular algorithms evaluate the integrand over a grid, whereas Monte Carlo methods randomly choose the points at which the integrand is evaluated. Keep in mind that similar to other numerical techniques, the Monte Carlo evaluation of integrals is only an approximation. Although Monte Carlo is efficient for multidimensional integrals, it is slow, requiring many samples for convergence.

References

1. C. W. Alexion et al., Evaluation of radiation fields using statistical methods of integration, *IEEE Trans. on Antennas and Propagation*, vol. AP-26, no. 2, 288–293, March 1979.
2. R. C. Millikan, The magic of the Monte Carlo method, *BYTE*, vol. 8, February 1988, pp. 371–373.
3. M. H. Kalos and P. A. Whitlook, *Monte Carlo Methods*, vol. 1, New York: John Wiley, 1986, pp. 89–116.
4. J. M. Hammersley and D. C. Handscomb, *Monte Carlo Methods*, London: Methuen, 1964.
5. G. Yang, A Monte Carlo method of integration, http://unicast.org/enclosures/text.pdf.
6. S. Haber, A modified Monte-Carlo quadrature II, *Math. Computation*, vol. 21, 388–397, July 1967.
7. S. Haber, Numerical evaluation of multiple integrals, *SIAM Review*, vol. 12, no. 4, October 1970, pp. 481–527.
8. J. M. Hammersley and K. W. Morton, A new Monte Carlo technique: Antithetic variates, *Proc. Cambridge Philosophical Society*, vol. 52, 1955, pp. 449–475.
9. J. H. Halton and D. C. Handsom, A method for increasing the efficiency of Monte Carlo integration, *J. ACM*, vol. 4, 329–340, 1957.
10. S. J. Yakowitz, *Computational Probability and Simulation*, Reading, MA: Addison-Wesley, 1977, pp. 192, 193.

Problems

11.1. Use sample mean Monte Carlo integration to evaluate:

a. $\displaystyle\int_0^1 4\sqrt{1-x^2}\,dx$

b. $\displaystyle\int_0^1 \sin x \, dx$

c. $\displaystyle\int_0^1 e^x dx$

d. $\displaystyle\int_0^1 \frac{1}{\sqrt{x}} dx$

11.2. Evaluate the following four-dimensional integrals:

a. $\displaystyle\int_0^1\int_0^1\int_0^1\int_0^1 \exp(x^1 x^2 x^3 x^4 - 1) dx^1 dx^2 dx^3 dx^4$

b. $\displaystyle\int_0^1\int_0^1\int_0^1\int_0^1 \sin(x^1 + x^2 + x^3 + x^4) dx^1 dx^2 dx^3 dx^4$

11.3. The radiation from a rectangular aperture with constant amplitude and phase distribution may be represented by the integral

$$I(\alpha, \beta) = \int_{-1/2}^{1/2} \int_{-1/2}^{1/2} e^{j(\alpha x + \beta y)} dx dy$$

Evaluate this integral using the Monte Carlo procedure, and compare your result for $\alpha = \beta = \pi$ with the exact solution

$$I(\alpha, \beta) = \frac{\sin(\alpha/2) \sin(\beta/2)}{\alpha \beta / 4}$$

Index

Milton Keynes UK
Ingram Content Group UK Ltd.
UKHW040103071024
449327UK00019B/777

9 780367 385804